Sergei Vladimirovich Bulanov

Lectures on Nonlinear Physics

APPUNTI

SCUOLA NORMALE SUPERIORE
2000

ISBN: 978-88-7642-267-6

Sergei Vladimirovich Bulanov
General Physics Institute
Russian Academy of Sciences
Vavilov Street, 38
Moscow, Russia
e-mail: bulanov@fpl.gpi.ru

Lectures on Nonlinear Physics

Contents

Preface

The main goal of these lectures is to make students get acquainted with a glossary of nonlinear physics. I try to present the material of the lectures in such a way that the students do not need to use additional literature during the first reading.

The lectures begin with the discussion of nonlinear problems in classical mechanics. Mainly I address to the explanation of how to use the multiple time scale expansion method to solve problems of nonlinear oscillations. Then the adiabatic invariants of slowly varying dynamic systems with periodic motion and the accuracy of their conservation are discussed. The anholonomy angles of slowly varying dynamic systems are considered. Generic properties of behavior of the dynamic systems near the separatrix trajectories in the phase plane in regard with the adiabatic invariant conservation and onset of the stochasticity are presented.

In the presentation of nonlinear properties of continuous media I start with the discussion of small amplitude wave propagation in dispersive media. The finite amplitude wave breaking plays the key role in an understanding of nonlinear wave propagation. Here I explain in detail how to describe nonlinear wave steepening and shock wave formation using the Lagrange variables. In unstable continuous systems nonlinear aspects of wave evolution acquire novel features. These features may be elucidated by considering the typical form of singularities formed at the nonlinear stage of instabilities of hydrodynamics type.

Then briefly I explain how to solve nonlinear elliptic equations considering the Liouville equation and its applications to describe the plasma equilibrium in the magnetic field and vortex systems in the fluid. Interacting point vortices in the Euler hydrodynamics and in the dispersive media provide an example of Hamiltonian systems with rich properties. I present an approach to investigating stability of the vortex systems and describe homogeneous vortex collapse in the systems with continuous and point distribution of the vorticity.

As in other modern courses on nonlinear physics the central part of these lectures is devoted to the soliton solutions of the Korteweg de Vries, Sine-Gordon and Nonlinear Schrödinger equation. I paid the most attention to the introduction into the inverse scattering method for the KdV equation. In the final lectures I explain how to use the Bäcklund transformation and analyze self-similarity of nonlinear systems to solve various nonlinear problems.

These notes are the outgrowth of a course of lectures with I have delivered at the Scuola Normale Superiore in Pisa during Aprile — June, 1999. I appreciate very much the scientists from SNS and students for their remarks and discussions. Special mention must be made of Professor Giuseppe Bertin and Professor Francesco Pegoraro for their encouragement and support throughout my lecturing.

Part I

Elements of Classical Mechanics

Chapter 1

Lagrange Mechanics

Lagrange mechanics. The principle of critical action. The Lagrange equations. The Neother theorem. Lagrange systems on the manifolds. The Jacoby integral. The Clairaut theorem.

***

1.1 The Principle of Critical Action

We shall describe the particle motion by the Lagrange equations, which appear naturally from the principle of critical action (Hamilton's principle):

Motion of a mechanical system is realized along the extremals of functional

$$\mathcal{S} = \int\limits_{t_2}^{t_1} \mathcal{L}dt, \tag{1.1}$$

where $\mathcal{L}(q_i, \dot{q}_i, t) = T - U$ is the Lagrange function. Here T and U are the kinetic and potential energy, respectively; q_i are the generalized coordinates, $\dot{q}_i$ are the generalized velocities; $\partial\mathcal{L}/\partial\dot{q}_i$ are the generalized momenta; $\partial\mathcal{L}/\partial q_i$ are the generalized forces; $\mathcal{S} = \int\limits_{t_2}^{t_1} \mathcal{L}(q_i, \dot{q}_i, t)dt$ is the action.

The trajectory of the particle motion $q_i = q_i(t)$ corresponds to the extremals of functional (1.1). To find it we consider a variation of functions $q_i(t)$ in the form

$$q_i(t) + \delta q_i(t), \tag{1.2}$$

with $\delta q_i(t)$ supposed to be small, and $\delta q_i(t_1) = \delta q_i(t_2) = 0$.

We calculate the variation of the action

$$\delta\mathcal{S} = \delta \int\limits_{t_2}^{t_1} \mathcal{L}(q_i, \dot{q}_i, t)dt = \int\limits_{t_2}^{t_1} \left(\frac{\partial\mathcal{L}}{\partial q_i}\delta q_i + \frac{\partial\mathcal{L}}{\partial \dot{q}_i}\delta\dot{q}_i \right) dt. \tag{1.3}$$

Since $\delta\dot{q}_i = \frac{d}{dt}\delta q_i$, integrating the second term into the right-hand side of the equation by part we obtain

$$\delta S = \left.\frac{\partial\mathcal{L}}{\partial\,\dot{q}_i}\delta q_i\right|_{t_1}^{t_2} + \int_{t_2}^{t_1}\left(\frac{d}{dt}\left(\frac{\partial\,\mathcal{L}}{\partial\dot{q}_i}\right) - \frac{\partial\mathcal{L}}{\partial q_i}\right)\delta q_i dt. \qquad (1.4)$$

Due to the condition $\delta q_i(t_1) = \delta q_i(t_2) = 0$ the first term in the right-hand side vanishes. According to the Hamilton principle $\delta S = 0$ we obtain the Euler-Lagrange equation

$$\frac{d}{dt}\left(\frac{\partial\mathcal{L}}{\partial\dot{q}_i}\right) - \frac{\partial\mathcal{L}}{\partial q_i} = 0, \qquad i = 1,\dots,n, \qquad (1.5)$$

for functions $q_i(t)$. In mechanics these equations are known as the Lagrange equations.

The Lagrange formalism provides a powerful tool for writing equations of motion in the systems with constraints. We consider the Lagrange systems with holonomic (integrable) constraints.

Example. The Lagrange function of the charged particle in the electric and magnetic fields, **E** and **B**, is

$$\mathcal{L} = -mc^2\left(1 - \frac{\mathbf{v}^2}{c^2}\right)^{1/2} + \frac{e}{c}\mathbf{A}\cdot\mathbf{v} - e\varphi. \qquad (1.6)$$

In nonrelativistic case, when $\mathbf{v}^2/c^2 \ll 1$, we have

$$\mathcal{L} = \frac{m\mathbf{v}^2}{2} + \frac{e}{c}\mathbf{A}\cdot\mathbf{v} - e\varphi. \qquad (1.7)$$

Here **A** and φ are the vector and scalar potentials. The electric and magnetic fields are expressed in the form: $\mathbf{E} = -c^{-1}\partial_t\mathbf{A} - \nabla\varphi$ and $\mathbf{B} = \nabla\times\mathbf{A}$.

•

1.1.1   The Noether Theorem

Now let us suppose that the action S is invariant under a group of transformations given by a group G_r and the infinitesimal operators which correspond to the group G_r are

$$X_\alpha = \xi_\alpha\frac{\partial}{\partial t} + \eta_\alpha^i\frac{\partial}{\partial q_i}, \qquad \alpha = 1,\dots,r. \qquad (1.8)$$

In this case in addition to the Euler equation we have independent conservation laws:

$$\frac{d}{dt}A_\alpha = 0, \qquad \alpha = 1,\dots,r, \qquad (1.9)$$

where functions $A_\alpha(t,q_i,\dot{q}_i)$ are given by expressions

$$A_\alpha = (\eta_\alpha^i - \dot{q}_i\xi_\alpha)\frac{\partial\mathcal{L}}{\partial\dot{q}_i} + \xi_\alpha\mathcal{L}. \qquad (1.10)$$

As an example we consider well known conservation laws in classical mechanics for the particle motion described by the Euler equations

$$m\ddot{x}_i = 0, \qquad i = 1, 2, 3, \tag{1.11}$$

where $x_i(t)$ are the particle coordinates. The Lagrange function is $\mathcal{L} = \sum_i m\dot{x}_i^2/2$. Writing the infinitesimal operators as

$$X = \xi\frac{\partial}{\partial t} + \eta^i\frac{\partial}{\partial x_i}, \tag{1.12}$$

we obtain the function $A(t, x_i, \dot{x}_i)$:

$$A = m\sum_i \dot{x}_i \left(\eta^i - \frac{1}{2}\xi\dot{x}_i \right). \tag{1.13}$$

Equations (1.11) admit the translation groups along the coordinates x_i, time t, and the group of rotation which are given by operators

$$X_i = \frac{\partial}{\partial x_i}, \qquad i = 1, 2, 3, \qquad X_4 = \frac{\partial}{\partial t},$$

$$X_{k,l} = x_l\frac{\partial}{\partial x_k} - x_k\frac{\partial}{\partial x_l}, \qquad k, l = 1, 2, 3. \tag{1.14}$$

From equations (1.11) we obtain conserved quantities. Invariance with respect to the translation along the coordinates x_i corresponds to $A_i = m\dot{x}_i$ which is the particle momentum $\mathbf{p}$. Translations in time give $A_4 = \sum_i m\dot{x}_i^2/2$ with the particle energy E. Invariance with respect to the rotation around the axes x_i corresponds to the rotation momentum conservation $\mathbf{M} = \mathbf{p} \times \mathbf{x}$.

1.2   Lagrange Systems on the Manifolds

We suppose that the particle can move only in the surface σ. To describe this motion we introduce the curvilinear coordinates $q_1, q_2, \ldots, q_n$. Here q_1 is the measure of the distance from the surface σ and $q_2, \ldots, q_n$ are the coordinates in the surface σ.

Motion in the surface is described by the Lagrange equations

$$\frac{d}{dt}\left(\frac{\partial \mathcal{L}^*}{\partial \dot{q}_j} \right) - \frac{\partial \mathcal{L}^*}{\partial q_j} = 0, \qquad j = 2, \ldots, n, \tag{1.15}$$

where

$$\mathcal{L}^*(q_j, \dot{q}_j, t) = \mathcal{L}(q_i, \dot{q}_i, t)|_{q_1 = \dot{q}_1 = 0}. \tag{1.16}$$

The Lagrange equations in the form (1.15) describe the system with the holonomic constraints. In a real mechanical system the constraints are due to some forces (we can assume them to be potential) that trap the particle close to the surface $q_1 = 0$. In some cases we must find the conditions under which the particle trajectory remains in the neighborhood of the surface $q_1 = 0$. To do that we need to find the solution of equations (1.15) and substitute dependencies $q_j(t)$, $\dot{q}_j(t)$ ($j = 2, \ldots, n$) into the expression for $\mathcal{L}(q_j, \dot{q}_j, t)$ to prove when equation

$$\frac{d}{dt}\left(\frac{\partial \mathcal{L}}{\partial \dot{q}_1}\right) - \frac{\partial \mathcal{L}}{\partial q_1} = 0 \tag{1.17}$$

can be fulfilled.

Example. We consider the motion of the particle with the mass, m, along the circle with the radius r. The circle rotates with the angular velocity ω around the vertical axis in the gravitation field $g\mathbf{e_z}$. As the coordinate q we take the angular coordinate along the circle. The Cartesian coordinates, x, y, z, are expressed via the coordinates, $r, q, \varphi = \omega t$, as

$$x = r \sin q \cos \omega t, \qquad y = r \sin q \sin \omega t, \qquad z = r \cos q. \tag{1.18}$$

From these formulas we find the kinetic energy,

$$T = \frac{m}{2}(\dot{x}^2 + \dot{y}^2 + \dot{z}^2) = \frac{m}{2}(\omega^2 r^2 \sin^2 q + r^2 \dot{q}^2), \tag{1.19}$$

and the potential energy,

$$U = mgr \cos q. \tag{1.20}$$

The Lagrange function $\mathcal{L} = T - U$,

$$\mathcal{L} = \frac{mr^2}{2}\dot{q}^2 + \frac{m\omega^2 r^2}{2}\sin^2 q - mgr \cos q, \tag{1.21}$$

is as same as in the case of the motion of the particle with the mass mr^2 in the field with the potential $V(q) = mgr \cos q - (m\omega^2 r^2/2) \sin^2 q$. When rotation is slow enough ($\omega^2 r < g$) the lower position on the circle is stable, and the particle oscillates around this position.

When we have $\omega^2 r > g$, the rotation is fast, the lower position becomes unstable and two new stable positions, for which $\cos q = -(g/\omega^2 r)$, appear.

This provides an example of bifurcation, when, as a result of the parameter ω change one stable solution disappears and instead two stable and one unstable solutions arise.

•

Exercise. Find frequencies of oscillations in both cases.

◇

Example. We consider motion of a particle on a parabola $y = kx^2$ rotating with a constant angular velocity Ω about the y−axis. For the present problem kinetic and potential energy equal $T = m(\dot{x}^2 + \Omega^2 x^2 + \dot{z}^2)/2$ and $U = mgz$. Using the constraint $y = kx^2$, we obtain the Lagrange function

$$\mathcal{L} = \frac{m}{2}\left((1 + 4k^2 x^2)\dot{x}^2 + \Omega^2 x^2\right) - mgkx^2. \tag{1.22}$$

The equation describing the motion of the particle is the following

$$(1 + 4k^2 x^2)\ddot{x} + (2gk - \Omega^2)x + 4k^2 \dot{x}^2 x = 0. \tag{1.23}$$

This equation has the integral

$$\left(1 + 4k^2 x^2\right) \dot{x}^2 + (2gk - \Omega^2)x^2 = h, \tag{1.24}$$

which is the Jacoby integral.

When $2gk > \Omega^2$, the motion is periodic with the trajectory confined in the neighborhood of the origin; while for $2gk < \Omega^2$ the motion is unbounded and the origin is a saddle point in the phase plane.

●

1.2.1   Jacoby Integral

We consider the Lagrangian independent of time. Multiplying equation (1.5) by $\dot{q}$ and adding and subtracting $\ddot{q}\partial\mathcal{L}/\partial\dot{q}$, we rewrite the equation as

$$\frac{d}{dt}\left(\frac{\partial\mathcal{L}}{\partial\dot{q}}\dot{q} - \mathcal{L}\right) = 0, \tag{1.25}$$

which is equivalent to conservation of

$$\frac{\partial\mathcal{L}}{\partial\dot{q}}\dot{q} - \mathcal{L} = h. \tag{1.26}$$

This is the Jacoby integral.

Example.   *The surfatron acceleration.* We consider the motion of a charged particle trapped by the potential wave in the magnetic field, when the electrostatic potential and the vector potential in the Lagrange function (1.6) are given by expressions

$$\varphi(x,t) = -\frac{E}{k}\sin(k(x - Vt)), \quad \text{and} \quad \mathbf{A} = Bz\mathbf{e}_x, \tag{1.27}$$

respectively. These expressions correspond to the electrostatic wave with the wavelength $2\pi/k$ propagating along the $x-$axis with the velocity V. The electric field in the wave, $\mathbf{E}$, is parallel to the x-axis. The homogeneous magnetic field, $\mathbf{B}$, is directed along the $y-$ axis. The Lagrange function (1.6) reads

$$\mathcal{L} = -mc^2\left(1 - \frac{\dot{x}^2}{c^2} - \frac{\dot{z}^2}{c^2}\right)^{1/2} + \frac{e}{c}Bz\dot{x} + \frac{eE}{k}\sin(k(x - Vt)). \tag{1.28}$$

We assume that the particle trajectory is localized in the plane $x = X + Vt$. That is $\dot{x} = V$. Using this as the constraint we write the Lagrange function $\mathcal{L}^*$ as

$$\mathcal{L}^* = -mc^2\left(1 - \beta^2 - \frac{\dot{z}^2}{c^2}\right)^{1/2} + e\beta Bz, \tag{1.29}$$

where $\beta = V/c$. From the Lagrange equation (1.15) with the Lagrangian (1.29) we find that the particle energy $E = mc^2\left(1 - \beta^2 - \dot{z}^2/c^2\right)^{-1/2}$ depends on time as

$$E = \left(\frac{1 + (\omega_B\beta t)^2}{1 - \beta^2}\right)^{1/2} \tag{1.30}$$

with $\omega_B = eB/mc$ being the Larmor frequency. The z-component of the particle velocity depends on time as

$$\dot{z} = \omega_B \beta t \left(\frac{1 - \beta^2}{1 + (\omega_B \beta t)^2} \right)^{1/2}. \tag{1.31}$$

To find the conditions of the particle trapping in the moving potential well we insert (1.31) into (1.28) and obtain that for $t \to \infty$ the condition

$$E \geq \frac{B}{\left(1 - \beta^2\right)^{1/2}} \tag{1.32}$$

must be fulfilled.

This is the surfatron acceleration which is also called as the $\mathbf{v} \times \mathbf{B}$ acceleration (*T. Katsouleas and J. Dawson*). It has been suggested as a mechanism of a charged particle acceleration in the laser-plasma interaction and invoked to describe basic properties of particle acceleration by the collisionless shock waves in magnetized plasmas.

●

Example. *Motion along the curve front.* Now we assume that the shock wave front can be approximated by a smooth surface, whose shape and position are determined by the given function of time $S(\mathbf{r}, t) = 0$. We consider the shock wave with the front of the form of a rotation surface described by the expression

$$r = r_f(z, t) = Vt - \frac{1}{2}\alpha(t)z^2 + \dots . \tag{1.33}$$

Here α is a curvature in the plane $\varphi =$ constant. The magnetic field is supposed to be directed along the z-axis; the vector potential has only the φ-component

$$\mathbf{A} = \mathbf{e}_\varphi \int^r B_z(r)r dr. \tag{1.34}$$

The Lagrange function $\mathcal{L}^*$ that corresponds to our choice of the form of the shock wave front can be written as follows

$$\mathcal{L}^*(z, \varphi, \dot{z}, \dot{\varphi}) = \frac{e\dot{\varphi}}{c} \int\limits_0^{r_f(z,t)} B_z(r)r dr +$$

$$-mc^2 \left(1 - \frac{\dot{r}_f^2 + 2\dot{r}_f r_f' \dot{z} + r_f^2 \dot{\varphi}^2(1 + (r_f')^2)\dot{z}^2}{c^2} \right)^{1/2}. \tag{1.35}$$

Here the prime denotes a differentiation with respect to the z-coordinate.

In the next step of our analysis it is necessary to pay attention to the fact that the Lagrangian does not depend on the φ-coordinate so that the φ-component of the generalized momentum is conserved according to the Noether theorem:

$$P_\varphi = r_f^2(\gamma\dot{\varphi} + \Omega(r_f)) = \text{const.} \tag{1.36}$$

Here we have introduced a function

$$\Omega(r_f) = \frac{e}{mcr_f^2} \int\limits_0^{r_f(z,t)} B_z(r)rdr.$$

(1.37)

Let us notice that in the uniform magnetic field the frequency $\Omega(r_f) = eB/2mc$ is equal to one half of the Larmor frequency.

Now we assume that the radius of the shock wave front depends on time as $r_f(z,t) = Vt$. The trajectory that corresponds to the initial conditions $z(0) = 0$ and $\dot{z}(0) = 0$, lies in the equatorial plane, $z = 0$. In the limit $t \to \infty$ the particle orbit is given by relationships

$$\dot{\varphi} = -\frac{\Omega(r_f)}{\gamma}, \qquad \text{and} \qquad \gamma = \left(\frac{c^2 + \Omega^2(r_f)r_f^2}{c^2 - V^2} \right)^{1/2}$$

(1.38)

When $B_z = $ const we have for $t \to \infty$

$$\gamma \approx \frac{eB_z r_f}{2(1 - \beta^2)^{1/2}mc^2},$$

(1.39)

and in the ultrarelativistic limit the trajectory has the form of a logarithmic spiral given by expression

$$\varphi = \frac{(1 - \beta^2)^{1/2}}{\beta} \ln r,$$

(1.40)

while in the nonrelativistic case it has the form of the Archimedes spiral:

$$\varphi = \frac{eB_z}{mcV}r.$$

(1.41)

Let us consider the form of the trajectory when $z(0) \neq 0$ and/or $\dot{z}(0) \neq 0$, assuming that perturbations are small. Linearizing the equations of motion for the z−coordinate of the particle we obtain that the motion along the z-axis for small displacement $z^{(1)}$ is described by

$$\frac{d}{dt}\left(\gamma^{(0)} \frac{dz^{(1)}}{dt} \right) - \alpha \frac{\Omega^2(r_f)Vt}{\gamma^{(0)}} \left(1 + \frac{d\ln\Omega(r_f)}{d\ln r_f} \right) z^{(1)} = 0.$$

(1.42)

Using equation (1.39) for the Lorenz factor $\gamma^{(0)}$ and assuming that Ω does not depend on r_f we rewrite the equation (1.42) in the ultrarelativistic limit (for $\gamma^{(0)} \gg 1$) as

$$\ddot{z}^{(1)} + \frac{1}{t}\dot{z}^{(1)} - \alpha \frac{(1 - \beta^2)c}{\beta t} z^{(1)} = 0,$$

(1.43)

and in the nonrelativistic (for $\gamma^{(0)} \ll 1$) limit as

$$\ddot{z}^{(1)} - \alpha\Omega^2 Vt z^{(1)} = 0.$$

(1.44)

We see that stability of the particle motion is determined by the sign of the curvature, α. Using the WKB method (see below) we find that asymptotically in the limit $t \to \infty$ in the ultrarelativistic case

$$z^{(1)} \sim (t_1/t)^{1/4} \cos((t/t_1)^{1/2}), \qquad \text{for} \qquad \alpha < 0, \qquad (1.45)$$

$$z^{(1)} \sim (|t_1|/t)^{1/4} \exp((t/|t_1|)^{1/2}), \qquad \text{for} \qquad \alpha > 0, \qquad (1.46)$$

where $t_1 = -\beta/4\alpha c(1 - \beta^2)$, and in the nonrelativistic ($\gamma^{(0)} \ll 1$) case

$$z^{(1)} \sim (t_0/t)^{1/4} \cos((t/t_0)^{3/2}), \qquad \text{for} \qquad \alpha < 0, \qquad (1.47)$$

$$z^{(1)} \sim (|t_0|/t)^{1/4} \exp((t/|t_0|)^{3/2}), \qquad \text{for} \qquad \alpha > 0, \qquad (1.48)$$

where $t_0 = (9/\alpha\Omega^2 V)^{1/3}$.

If we have an expanding shock wave with the front curvature decreasing with time as $\alpha \sim 1/Vt$, we find that

$$z^{(1)} \sim t^{(1-\beta)/\beta} \qquad \text{for} \qquad \alpha > 0 \qquad (1.49)$$

in the ultrarelativistic case, and

$$z^{(1)} \sim \exp(\Omega t) \qquad (1.50)$$

in the nonrelativistic case.

To obtain the conditions of the particle trapping at the shock front we substitute the above-found expressions (1.38) into the Euler-Lagrange equation for the r-component. When $\mathbf{B} = B_z \mathbf{e}_z$, the equation for the r-coordinate of the particle can be written as

$$\frac{d}{dt}\left(\gamma \frac{dr}{dt}\right) = \gamma r \left(\frac{d\varphi}{dt}\right)^2 + \frac{eB_z}{mc} r \frac{d\varphi}{dt} + \frac{e}{m} E(r). \qquad (1.51)$$

Here the electric field in the wave $E = -\partial\phi/\partial r$, and the electrostatic potential can be approximated as

$$\phi(r,t) = \frac{E_0}{(r\kappa)^{1/2}} \sin(\kappa(r - Vt) + \epsilon) = E_m \sin(\kappa(r - Vt) + \epsilon). \qquad (1.52)$$

Further we neglect slow dependence of the electric field amplitude on time assuming E_m to be constant.

Substituting dependence $\dot{\varphi}$ from equation (1.38) into equation (1.51) we obtain

$$\frac{d}{dt}\left(\gamma \frac{dr}{dt}\right) = -\frac{1}{\gamma}\left(\frac{eB_z}{2mc}\right)^2 r + \frac{e}{m} E(r,t). \qquad (1.53)$$

The particle trapped by the shock wave has the radial velocity equal V. Equation (1.53) has a solution $r(t) = Vt$ for

$$E = E_m \sin \epsilon = \frac{B_z}{2(1 - \beta^2)^{1/2}}. \qquad (1.54)$$

Since $|\sin \epsilon| \leq 1$, we find the condition of unlimited acceleration

$$\frac{E_m}{B_z} > \frac{1}{2(1 - \beta^2)^{1/2}}, \tag{1.55}$$

which is in a factor of two larger than the condition (1.32) found above.

•

1.2.2   The Clairaut Theorem

If we suppose that the radius of the shock wave front, $r_f(z)$, does not change in time, the particle energy, $E = \gamma mc^2$, is conserved. Here γ is the Lorentz factor. In this case the absolute value of the particle velocity, $|\mathbf{v}|$, is constant. We introduce the angle ψ between the velocity direction and the direction of the meridian on the surface $r_f(z)$. Then we take into account that $r_f(z)\dot{\varphi} = |\mathbf{v}| \sin \psi$. From equation (1.36) we obtain

$$\gamma r_f(z)|\mathbf{v}| \sin \psi + r_f^2 \Omega(r_f) = P_\varphi. \tag{1.56}$$

When the magnetic field vanishes, $\Omega(r_f) = 0$, we have the relationship

$$r_f(z) \sin \psi = \text{const.} \tag{1.57}$$

This is equivalent to the Clairaut theorem. It shows that the particle trajectory is confined within the region $|\sin \psi| \leq 1$, i.e. for $r \geq r_0 \sin \psi_0$. In the case when the magnetic field action is taken into account the trajectory lies in the region determined by expression

$$r = -\frac{\gamma|\mathbf{v}| \sin \psi}{2\Omega} \pm \left(\left(\frac{\gamma|\mathbf{v}| \sin \psi}{2\Omega} \right)^2 + \frac{2P_\varphi}{\Omega} \right)^{1/2}. \tag{1.58}$$

Chapter 2

Hamilton Mechanics

Hamilton Mechanics. Legendre transform. Hamilton equations. Hamiltonian flux in the phase space. The Liouville theorem. Canonical transformations. Symplectic conditions. The Hamiltonian approach in electrodynamics. The U, V-canonical transformation for the harmonic oscillator. Hamiltonization of a system of ordinary differential equations.

***

By the Legendre transform the Lagrange system can be transformed into the Hamilton system.

2.1 The Legendre Transform

The Legendre transform changes a function $y = f(x)$ with $y''(x) > 0$ into a new function g of a new variable p:

$$g(p) = (px - f(x))|_{x=x(p)}, \qquad (2.1)$$

where $x(p)$ must be found from equation $f'(x) = p$.

Exercise. Demonstrate that

$$f(x) = (px - g(p))|_{p=p(x)}, \qquad (2.2)$$

where $p(x)$ must be found from equation $g'(p) = x$.

∞

Exercise. Demonstrate that

$$px \leq f(x) + g(p) \qquad (2.3)$$

for $f(x)$ and $g(p)$ are Legendre transforms each of other.

∞

The Hamilton function $\mathcal{H}(p_i, q_i, t)$ is equal to the Legendre transform of the Lagrange function to new variables $p_i = \partial L/\partial \dot{q}_i$, q_i:

$$\mathcal{H}(p_i, q_i, t) = p_i \dot{q}_i - \mathcal{L}(q_i, \dot{q}_i). \tag{2.4}$$

The equation of motion takes the form of Hamilton equations:

$$\dot{p}_i = -\frac{\partial \mathcal{H}}{\partial q_i}, \qquad \dot{q}_i = \frac{\partial \mathcal{H}}{\partial p_i}. \tag{2.5}$$

The Hamiltonian systems have important properties of conservation of the Hamiltonian value for the Hamiltonian independent explicitly of time

$$\mathcal{H}(p_i(t), q_i(t)) = \mathcal{H}(p_i(0), q_i(0)) = h, \tag{2.6}$$

and conservation of the phase volume during motion of the particle ensemble (the Liouville theorem).

The Liouville theorem is equivalent to the condition that the Jacobian of the transformation from the coordinates $p_i(0), q_i(0)$ to $p_i(t), q_i(t)$ is conserved

$$D = \frac{\partial(\mathbf{p}(t), \mathbf{q}(t))}{\partial(\mathbf{p}(0), \mathbf{q}(0))} = 1. \tag{2.7}$$

It means that the motion of the particle ensemble in the phase space is similar to motion of incompressible fluid.

A canonical transformation from the coordinates $p_i(t), q_i(t)$ to $P_i(t), Q_i(t)$, which obey equations with the Hamiltonian $\widetilde{H}\,(\mathbf{P}, \mathbf{Q}, t)$, is given by the generation function $\mathcal{S}(\mathbf{q}, \mathbf{Q}, t)$ with

$$p_i = \frac{\partial \mathcal{S}}{\partial q_i}, \qquad P_i = \frac{\partial \mathcal{S}}{\partial Q_i}, \qquad \text{and} \qquad \widetilde{H} = \mathcal{H} + \frac{\partial \mathcal{S}}{\partial t}. \tag{2.8}$$

Example. The Hamilton function of a charged particle in the electric and magnetic field, $\mathbf{E}$ and $\mathbf{B}$, is

$$\mathcal{H} = \left(\left(\mathbf{p} - \frac{e}{c}\mathbf{A} \right)^2 c^2 + m^2 c^4 \right)^{1/2} + e\varphi. \tag{2.9}$$

●

Example.   The magnetic field $\mathbf{B}$ is a divergence-free (solenoidal) vector field, the magnetic field line flow is incompressible. That is why it is convenient to write the magnetic field structure in a plasma in a Hamiltonian form. The simplest configuration corresponds to the magnetic field that depends on two coordinates x, y and has the form

$$\mathbf{B}(x, y) = \nabla \times (A_z \mathbf{e}_z) + B_z \mathbf{e}_z, \tag{2.10}$$

where $A_z(x, y)$ is the $z-$component of the vector potential, $\mathbf{e}_z = \nabla z$, is a unit vector in the $z-$direction, and the $z-$ component of the magnetic field $B_z(x, y)$ is assumed to be independent of z. The field-line equation reads

$$\frac{dx}{B_x} = \frac{dy}{B_y} = \frac{dz}{B_z} = ds, \tag{2.11}$$

where s is the parameter along the field line. Using $B_x = \partial_y A_z$ and $B_y = -\partial_x A_z$, the projection of the field line onto a plane $z =$ constant is represented by equations of motion in a Hamiltonian form

$$\frac{dx}{ds} = \frac{\partial A_z}{\partial y}, \qquad \frac{dy}{ds} = -\frac{\partial A_z}{\partial x}. \tag{2.12}$$

Here $A_z(x, y)$ is the Hamiltonian. Because the Hamiltonian is a constant of motion $(\partial A_z/\partial s = 0)$ and the space dimension is one, the field-line equations are integrable, and every field line is on the surface $A_z(x, y) =$ constant.

If we assume that $\mathbf{B}$ has the toroidal symmetry $(\partial_\varphi \mathbf{B} = 0$; φ is the toroidal angle) we can express the magnetic field via two scalar functions $\psi(r, z)$ and $B_\varphi(r, z)$ as

$$\mathbf{B}(r, z) = \nabla \psi \times \mathbf{e}_\varphi + B_\varphi \mathbf{e}_\varphi, \tag{2.13}$$

where r, z, φ are the cylindrical coordinates. Using the magnetic field line equations $(d\mathbf{r}/\mathbf{B} = ds)$ we obtain that the projection of the field line onto a plane $\varphi =$ constant is represented by Hamiltonian equations

$$\frac{dr}{ds} = \frac{\partial \psi}{\partial z}, \qquad \frac{dz}{ds} = -\frac{\partial \psi}{\partial r}. \tag{2.14}$$

In this case $\psi(r, z)$ is the Hamiltonian, it is a constant of motion $(\partial_s \psi = 0)$ and the space dimension is one, the field-line equations are integrable, and every field line is on the surface $\psi(r, z) =$ constant. However it is not so if one does not assume a symmetry; in this case the general field-line structure is very complicated.

•

2.2   Hamiltonian Flux in the Phase Space

We assume that in the phase space the antisymmetric tensor K_{ij}: $K_{ij} = -K_{ji}$, is given. If

$$\frac{\partial K_{ij}}{\partial x_k} + \frac{\partial K_{jk}}{\partial x_i} + \frac{\partial K_{ki}}{\partial x_j} = 0 \tag{2.15}$$

and

$$\det K_{ij} \neq 0, \tag{2.16}$$

the space is symplectic and has even dimension; otherwise $\det K_{ij} = 0$. Here $x_i = (p_i, q_i)$ are the coordinates in the phase space. The system of ordinary differential equations is Hamiltonian if on the phase space a function $\mathcal{H}$ is given and

$$K_{ij}\dot{x}_j = \frac{\partial \mathcal{H}}{\partial x_i}. \tag{2.17}$$

In the simply connected domain equations (2.15) can be solved. The solution is

$$K_{ij} = \frac{\partial A_i}{\partial x_j} - \frac{\partial A_j}{\partial x_i}, \tag{2.18}$$

where A_i is the "potential".

The Hamilton principle $\delta S = 0$ in this case corresponds to

$$\delta S = \delta \int_{t_2}^{t_1} (A_i \dot{x}_i + \mathcal{H})\, dt = 0. \tag{2.19}$$

Since $\det K_{ij} \neq 0$, one can find a tensor $J_{ij} = K_{ij}^{-1}$ which is inverse to the tensor K_{ij}. The Hamilton equations (2.5) read

$$\dot{x}_i = J_{ij} \frac{\partial \mathcal{H}}{\partial x_j}. \tag{2.20}$$

For the tensor J_{jm} the condition (2.15) is equivalent to

$$J_{km} \frac{\partial J_{ij}}{\partial x_m} + J_{im} \frac{\partial J_{jk}}{\partial x_m} + J_{jm} \frac{\partial J_{ki}}{\partial x_m} = 0. \tag{2.21}$$

The Poisson brackets read

$$\{A, B\} = j_{ij} \frac{\partial A}{\partial x_i} \frac{\partial B}{\partial x_j}. \tag{2.22}$$

By virtue of $J_{ij} = -J_{ji}$ we have $\{A, B\} = -\{B, A\}$ and $\{\{A, B\}, C\} + \{\{C, A\}, B\} + \{\{B, C\}, A\} = 0$. The latter is Jacobi's identity.

Equations (2.20) describe a flux of the "fluid" in the phase space:

$$\dot{x}_i = V_i(\mathbf{x}, t). \tag{2.23}$$

According to these equations the elementary volume of the phase space moves from its initial position x_i^0 at $t = 0$ (the Lagrange coordinate) to $x_i(t)$ (the Euler coordinate). For infinitesimal displacement from the initial position δx_i we have

$$x_i(t) = x_i^0 + \delta x_i(t).$$

The Jacobi matrix of the transformation ("the matrix of deformation in the phase space") is equal to

$$M_{ij} = \frac{\partial x_i}{\partial x_j^0}. \tag{2.24}$$

From equation (2.23) we have

$$\delta \dot{x}_i = w_{ij} \delta x_i. \tag{2.25}$$

Exercise. Show that $w_{ij} = \dot{M}_{ik} M_{kj}^{-1}$.

∞

The determinant of M_{ij} is a Jacobian of the transformation: $D = \det M_{ij}$. Since

$$\dot{D}/ D = \dot{M}_{ij} M_{ji}^{-1} = w_{ii} \equiv \operatorname{div} \mathbf{V} \tag{2.26}$$

and according to equations (2.20) $\mathrm{div}\mathbf{V} = 0$, the density of the "fluid" in the phase space does not change (the Liouville theorem).

Equations (2.5) have the form of the flux equations (2.20) with the matrix J_{ij} equal

$$J_{ij} = \begin{pmatrix} 0_{kl} & -\delta_{kl} \\ \delta_{kl} & 0_{kl} \end{pmatrix}, \tag{2.27}$$

where 0_{kl} and δ_{kl} are the zero and the unit Kronecker tensor, respectively, with $1 \leq k, l \leq N$. It easy to show that $J_{ik}J_{kj} = -\delta_{ij}$. Thus, the matrix M_{ij} is a symplectic matrix.

However the Liouville theorem is equivalent to the symplectic conditions only in the case of the motion of the system with a single degree of freedom. In a generic case symplectic conditions correspond to preserving the area which is calculated in the phase space as it follows. We consider three trajectories in the phase space, $x_i(t)$, $x_i(t) + \delta x_i(t)$ and $x_i(t) + \delta x_i'(t)$, separated by infinitesimal distances. The symplectic differential area

$$\sigma = J_{ij}\delta x_i(t)\delta x_i'(t) = J_{ij}M_{ik}M_{jl}\delta x_k^0 \delta x_l^{0\prime} \tag{2.28}$$

is independent of time. To show this we differentiate $J_{ij}M_{ik}M_{jl}$ with respect to time. We write the relationships in the matrix form:

$$\dot{\hat{M}}^\top \hat{J}\hat{M} + \hat{M}^\top \hat{J} \dot{\hat{M}} = \hat{M}^\top((\dot{\hat{M}} \hat{M}^{-1})^{-1} \hat{J} + \hat{J} \dot{\hat{M}} \hat{M}^{-1})\hat{M}^{-1}$$

$$= \hat{M}^\top (\hat{H}^\top \hat{J}^\top \hat{J} + \hat{J}\hat{J}\hat{H})\hat{M} = 0, \tag{2.29}$$

where $H_{ij} = \partial^2 \mathcal{H}/\partial x_i \partial x_j$. We have used the relationships: $J^\top J = I$, $JJ = -I$, and $J^\top = -J$.

Now we consider the transformation from the coordinates $\mathbf{x} = (\mathbf{p}, \mathbf{q})$ to $\mathbf{X} = (\mathbf{P}, \mathbf{Q})$. Calculating the Jacobi matrix $M_{ij} = \partial X_i/\partial x_j$ and $\dot{\mathbf{X}}$, we find that

$$\dot{X}_i = M_{ij}J_{jm}M_{km}\frac{\partial \tilde{H}}{\partial X_k} = \tilde{J}_{ij}\frac{\partial \tilde{H}}{\partial X_j}, \tag{2.30}$$

where $\tilde{J}_{ij} = M_{ik}J_{kl}M_{jl}$. The equations are Hamiltonian if $\tilde{J}_{ij} = M_{ik}J_{kl}M_{jl} = J_{ij}$. The transformation is canonical if

$$M_{ik}J_{kl}M_{jl} = J_{ij}. \tag{2.31}$$

Definition. If for a matrix M_{ij} one has the relationship $M_{ik}J_{kl}M_{jl} = J_{ij}$ with J_{ij} given by (2.27), then M_{ij} is symplectic.

★

Since a unitary matrix is symplectic, the product of symplectic matrices is symplectic and the inverse symplectic matrix is symplectic, they form the symplectic group.

2.2.1   The Hamiltonian Approach in Electrodynamics

We describe the electromagnetic field in terms of the vector potential $\mathbf{A}(\mathbf{r},t)$. We represent the vector potential in the form

$$\mathbf{A}(\mathbf{r},t) = \sum_{\mathbf{k}} \mathbf{A}(\mathbf{k},t) \exp(i\mathbf{k} \cdot \mathbf{r}). \tag{2.32}$$

The vector potential is a real function; that is why we have $\mathbf{A}(\mathbf{k},t) = \mathbf{A}^*(-\mathbf{k},t)$, where " $*$ " denotes a complex conjugation. By virtue of the condition $\operatorname{div}\mathbf{A} = 0$ we have $(\mathbf{k} \cdot \mathbf{A}(\mathbf{k},t)) = 0$.

Assuming that the electromagnetic field is inside a box with the sides A, B, and C and with a volume $V = ABC$ we obtain $\Delta n = Vk^2 dk dO / (2\pi)^3$. From the Maxwell equations we have

$$\ddot{\mathbf{A}} + k^2 c^2 \mathbf{A} = 0. \tag{2.33}$$

The energy of the electromagnetic field is equal to

$$E = \frac{1}{8\pi} \int dV \left(|\mathbf{E}|^2 + |\mathbf{B}|^2 \right), \tag{2.34}$$

where $\mathbf{E} = -\partial_t \mathbf{A}/c$, and $\mathbf{B} = i\mathbf{k} \times \mathbf{A}$. Substituting these relationships into equation (2.34) we obtain

$$E = \frac{V}{8\pi c^2} \sum_{\mathbf{k}} \left(\dot{\mathbf{A}}(\mathbf{k},t) \cdot \dot{\mathbf{A}}^*(\mathbf{k},t) + k^2 c^2 \mathbf{A}(\mathbf{k},t) \cdot \mathbf{A}^*(\mathbf{k},t) \right). \tag{2.35}$$

Writing $\mathbf{A}(\mathbf{k},t) = \mathbf{a}(\mathbf{k},t) - \mathbf{a}^*(-\mathbf{k},t)$ we obtain

$$\mathbf{A}(\mathbf{r},t) = \sum_{\mathbf{k}} \left(\mathbf{a}(\mathbf{k},t) \exp(i\mathbf{k} \cdot \mathbf{r}) + \mathbf{a}(-\mathbf{k},t)^* \exp(-i\mathbf{k} \cdot \mathbf{r}) \right). \tag{2.36}$$

Here each function depends on $(\mathbf{k} \cdot \mathbf{r} - \omega_0 t)$. That means that they are propagating waves.

Since $\dot{\mathbf{A}}(\mathbf{k},t) = ik(\mathbf{a}(\mathbf{k},t) - \mathbf{a}^*(-\mathbf{k},t))$, the energy E is the Hamilton function, it is

$$E = \mathcal{H} = \sum_{\mathbf{k}} E(\mathbf{k}) = \sum_{\mathbf{k}} \frac{Vk^2}{2\pi} \left(\mathbf{a}(\mathbf{k},t) \cdot \mathbf{a}^*(-\mathbf{k},t) \right) = \sum_{\mathbf{k}} \omega(\mathbf{k}) J(\mathbf{k}). \tag{2.37}$$

The function $J(\mathbf{k})$ corresponds to the boson number operator in the quantum field theory. Moreover, this is a form of the Hamiltonian written in the canonical coordinates, the action $J(\mathbf{k})$, and the angle $\theta(\mathbf{k}) = \omega(\mathbf{k})t + \text{const}$.

With the canonical transform

$$\mathbf{Q} = \left(\frac{V}{4\pi c^2} \right)^{1/2} (\mathbf{a} + \mathbf{a}^*), \qquad \mathbf{P} = -i\omega(\mathbf{k}) \left(\frac{V}{4\pi c^2} \right)^{1/2} (\mathbf{a} - \mathbf{a}^*) \tag{2.38}$$

the Hamiltonian (2.37) takes the form

$$\mathcal{H} = \sum_{\mathbf{k}} \mathcal{H}(\mathbf{P}, \mathbf{Q}; \mathbf{k}) = \sum_{\mathbf{k}} \left(|\mathbf{P}(t; \mathbf{k})|^2 + \omega^2(\mathbf{k})|\mathbf{Q}(t; \mathbf{k})|^2 \right). \tag{2.39}$$

The Hamilton equations

$$\dot{\mathbf{Q}} = \frac{\partial \mathcal{H}}{\partial \mathbf{P}}, \qquad \dot{\mathbf{P}} = -\frac{\partial \mathcal{H}}{\partial \mathbf{Q}} \tag{2.40}$$

give

$$\ddot{\mathbf{Q}}_{\mathbf{k}} + \omega_{\mathbf{k}}^2 \mathbf{Q}_{\mathbf{k}} = 0 \tag{2.41}$$

with $\omega_{\mathbf{k}}^2 = k^2 c^2$.

2.2.2 The U,V-canonical Transformation for the Harmonic Oscillator

For the harmonic oscillator with the Hamiltonian $\mathcal{H} = (p^2 + \omega_0^2 q^2)/2$ we perform transformation from the variables p, q to new variables a, a^*. The transformation is supposed to be linear, that is why it is given by the 2×2 matrix M_{ij}: $X_i = M_{ij} x_j$. The matrix M_{ij} we chose to be

$$M_{ij} = \begin{pmatrix} iv^* & -iu^* \\ -iv & iu \end{pmatrix}. \tag{2.42}$$

Here the complex functions u and v must satisfy the condition

$$uv^* - u^* v = -i \tag{2.43}$$

for the U, V-transformation to be canonical. In this case we have

$$q = \frac{1}{(2\omega_0)^{1/2}} (a + a^*), \qquad p = i \left(\frac{\omega_0}{2} \right)^{1/2} (a - a^*), \tag{2.44}$$

with the Hamiltonian $\mathcal{H} = \omega_0 a a^*$ and Hamilton equations

$$\dot{a} = i \frac{\partial \mathcal{H}}{\partial a^*}, \qquad \dot{a}^* = -i \frac{\partial \mathcal{H}}{\partial a}. \tag{2.45}$$

The U, V-transformation provides an example of a canonical transformation with the valency $c = i$. The canonical transformation has the valency c if instead (2.31) one has

$$M_{ik} J_{kl} M_{jl} = c J_{ij}. \tag{2.46}$$

2.3   Hamiltonization

Any system of ordinary differential equations can be transformed into the canonical form *(P.A.M. Dirac, 1958)*. To demonstrate that we consider the system of equations in the form

$$\dot{x}_i = f_i(\mathbf{x},t,\varepsilon) \qquad (i = 1,\ldots,n).$$ (2.47)

Now we define the canonical conjugate to $\mathbf{x}$ vector $\mathbf{y}$ of generalized momenta $(y_i; i = 1,\ldots,n)$ and the Hamiltonian

$$\mathcal{H}(\mathbf{x},\mathbf{y},t) = f_i(\mathbf{x},t,\varepsilon)\, y_i.$$ (2.48)

Equations of motion are

$$\dot{x}_i = \frac{\partial \mathcal{H}}{\partial y_i} \equiv f_i(\mathbf{x},t,\varepsilon), \qquad \dot{y}_i = -\frac{\partial \mathcal{H}}{\partial x_i} \equiv -\frac{\partial f_j(x,t,\varepsilon)y_j}{\partial x_i}.$$ (2.49)

When one investigates concrete systems of differential equations of low order it is more convenient not to perform the hamiltonization. However, in the case of high-order systems it is more easy to operate with one function $\mathcal{H}(x_i, y_i, t)$ even depending on twice higher number of variables.

Chapter 3

Nonlinear Oscillations

Small amplitude oscillations. Nonlinear oscillations. Method of multiple scales to solve the linear and nonlinear oscillation problems. Nonlinear shift of frequency. Nonlinear Resonance. Limit circle. The Van der Pol equation. Linear oscillator with a slowly varying frequency. WKB approximation. The parametric resonance. Motion in the high frequency field. Motion of systems with two degrees of freedom.

***

We consider the system motion in the vicinity of the equilibrium position $\mathbf{q}_0, \dot{\mathbf{q}}_0$. Since it is a critical point, i.e. $\partial U/\partial \mathbf{q}|_{\mathbf{q}_0} = 0$, we expand the Lagrangian in $\mathbf{x} = \mathbf{q} - \mathbf{q}_0$. This yields

$$\mathcal{L} = \frac{1}{2}a_{ij}\dot{x}_i\dot{x}_j - \frac{1}{2}b_{ij}x_ix_j + \mathcal{L}_1(\mathbf{x}, \dot{\mathbf{x}}), \qquad b_{ij} = \left.\frac{\partial^2 U}{\partial x_i \partial x_j}\right|_{\mathbf{x}=0}. \tag{3.1}$$

Here $\mathcal{L}_1(\dot{\mathbf{x}}, \mathbf{x})$ describes higher order terms.

3.1 Small Amplitude Oscillations

If we neglect higher than second order terms in equation (3.1) and assume that the matrices a_{ij} and b_{ij} are not degenerate we can diagonalize both a_{ij} and b_{ij} simultaneously. In this case we obtain a system of non-interacting linear oscillators. Each of them is described by equation

$$\ddot{x} + \omega_0^2 x = 0, \tag{3.2}$$

where $\omega_0^2 = \lambda$, with λ being the eigenvalue of the characteristic equation, $\det|a_{ij} - \lambda b_{ij}| = 0$. Depending on the sign of the eigenvalue we have three different cases. If $\lambda = \omega_0^2 > 0$, the solution of equation (3.2) is $x(t) = a\cos(\omega_0 t + \varphi)$, where the amplitude, a, and phase φ, depend on initial conditions. This solution describes oscillations near the equilibrium. If $\lambda = -\gamma^2 = 0$, the solution is $x(t) = at + b$ (indifferent equilibrium). In the case when $\lambda = \omega_0^2 < 0$, $x(t) = a \cosh(\gamma t + \varphi)$ (instability).

We modify the linear oscillator equation (3.2) to take into account the effects of dissipation with the damping rate, ν, and the driven force, which is assumed to be periodic with amplitude, f, and frequency Ω. It reads

$$\ddot{x} + 2\nu\dot{x} + \omega_0^2 x = f \cos \Omega t. \tag{3.3}$$

The Initial Problem for Linear Oscillator

We consider the case when the driven force vanishes, $f = 0$. We need to solve the ordinary differential equation

$$\ddot{x} + 2\nu\dot{x} + \omega_0^2 x = 0 \tag{3.4}$$

with initial conditions $x(0) = x_0$ and $\dot{x}(0) = \dot{x}_0$. The Laplace transform with imaginary variable $p = i\omega$ being applied to equation (3.4) for

$$x(\omega) = \int_0^\infty x(t) \exp(i\omega t) dt \tag{3.5}$$

results in

$$(\omega^2 + 2i\nu\omega - \omega_0^2)x(\omega) = \dot{x}_0 - i\omega x_0 \equiv g_0(\omega). \tag{3.6}$$

A solution to equation (3.4) is given by the inverse Laplace transform

$$x(t) = \frac{1}{2\pi} \oint_S \frac{g_0(\omega) \exp(-i\omega t) d\omega}{(\omega^2 + 2i\nu\omega - \omega_0^2)} =$$

$$\frac{1}{2\pi} \oint_S \frac{g_0(\omega) \exp(-i\omega t) d\omega}{(\omega - \omega_1)(\omega - \omega_2)} = \sum_{j=1,2} a_j \exp(-i\omega_j t). \tag{3.7}$$

The contours of integration, S, are chosen to fulfill the causality principle: $x(t) = 0$ for $t < 0$. Here $a_1 = g_0(\omega_1)/(\omega_1 - \omega_2)$ and $a_1 = g_0(\omega_2)/(\omega_2 - \omega_1)$.

Thus we have found that the linear oscillations can be described by functions

$$x = a \cos\theta, \quad \text{with} \quad \theta = \omega t + \varphi, \tag{3.8}$$

where a and φ are the amplitude and phase of oscillations. The frequency ω obeys the dispersion equation

$$\omega^2 + 2i\nu\omega - \omega_0^2 = 0. \tag{3.9}$$

Depending on the sign and magnitude of ν the solution describes slowly decaying ($\nu > 0, \nu \ll \omega_0$), or slowly growing ($\nu < 0, \nu \ll \omega_0$) oscillations, or a periodic motion when $\nu \gg \omega_0$.

Laplace Transformation. We assume that $f(t)$ is a function of real variable t,

$$f(t) = 0, \qquad \text{for} \qquad t < 0, \tag{3.10}$$

and

$$\int dt \exp(-\sigma t)|f(t)|dt < \infty, \tag{3.11}$$

where σ is some real number. One-sided Laplace transformation $\varphi(z)$ of function $f(t)$ is defined as

$$\varphi(z) = \int_0^\infty dt \exp(izt)f(t). \tag{3.12}$$

A variable z is supposed to be complex. The singular points of the function $\varphi(z)$ in the complex plane z are below the line $Im\{z\} = \sigma_0$, or, in other words, the function $\varphi(z)$ is a regular function throughout the half-plane $Im\{z\} > \sigma_0$. The inverse Laplace transformation is given by

$$f(t) = \frac{1}{2\pi} \oint_C dz \exp(izt)\varphi(z), \tag{3.13}$$

where a contour C is parallel to the axis $Re\{z\}$ and lies above the line $Im\{z\} = \sigma_0$. It is known that the Laplace transformation of a convolution of two functions, $f_1(t)$ and $f_2(t)$,

$$F(t) = (f_1 * f_2)(t) = \int_0^t d\tau f_1(\tau)f_2(t - \tau) \tag{3.14}$$

is equal to

$$\Phi(z) = \varphi_1(z)\varphi_2(z), \tag{3.15}$$

where $\varphi_1(z)$ and $\varphi_2(z)$ are the Laplace transformations of $f_1(t)$ and $f_2(t)$.

Linear Resonance.

In the case when the driven force is nonzero for $t \to \infty$ asymptotically we have an expression that describes driven oscillations

$$x = b\cos(\Omega t + \varphi) \tag{3.16}$$

with the amplitude

$$b = \frac{f}{\left((\omega_0^2 - \Omega^2)^2 + 4\nu^2\Omega^2\right)^{1/2}}, \tag{3.17}$$

and the phase

$$\varphi = \arctan\frac{2\nu\Omega}{\omega_0^2 - \Omega^2}. \tag{3.18}$$

The behavior of b as a function of Ω is shown in Fig. 3.1. for $\nu \ll \Omega, \omega_0$.

The amplitude of oscillations tends to f/ω_0 when $\Omega \to 0$; it reaches the maximum, $f/2\nu\omega_0$ for $\nu \ll \omega_0$, at the resonance where $\Omega = \omega_0$; and decreases as $\sim 1/\Omega^2$, when $\Omega \to \infty$.

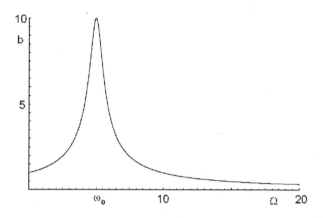

Fig. 3.1. Linear resonance.

3.2   Finite Amplitude Oscillations

3.2.1   Nonlinear Change of the Frequency of Free Oscillations

We discuss the effects of nonlinearity by using the Duffing equation as an example for anharmonic oscillator

$$\ddot{x} + \omega_0^2 x + \beta x^3 = 0. \tag{3.19}$$

The Duffing equation describes the material point motion under the action of the force with the potential

$$U(x) = \frac{\omega_0^2}{2} x^2 + \beta \frac{x^4}{4}. \tag{3.20}$$

This expression can be considered as first two terms in the expansion of generic dependence of the potential on the displacement x from the local minimum $x = 0$. Depending on the sign of the parameter β the potential has one minimum and grows when $x \to \infty$ when $\beta > 0$, as it is shown in Fig. 3.2 a, or it has the local minimum at $x = 0$ and two maximums at $x = \pm \left(\omega_0^2/\beta \right)^{1/2}$ for $\beta < 0$, as it is shown in Fig. 3.2 b.

We suppose the parameter β to be small and introduce small parameter $\varepsilon \ll 1$, which is in our case equal to

$$\varepsilon = \beta. \tag{3.21}$$

Straightforward expansion

Then we try to find a solution to the Duffing equation (3.19) expanding it into series on the small parameter ε :

$$x = x^{(0)} + \varepsilon x^{(1)} + \varepsilon^2 x^{(2)} + \dots. \tag{3.22}$$

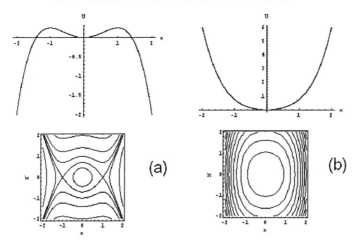

Fig. 3.2. Potential and phase plane: a) $\beta > 0$, b) $\beta < 0$.

To the zeroth order in ε we obtain equation

$$\ddot{x}^{(0)} + \omega_0^2 x^{(0)} = 0 \tag{3.23}$$

with the solution

$$x^{(0)} = a\cos\left(\omega_0 t + \varphi\right). \tag{3.24}$$

The first order in ε gives

$$\ddot{x}^{(1)} + \omega_0^2 x^{(1)} = -\left(x^{(0)}\right)^3 =$$

$$-\left(\frac{a^3}{4}\left(\cos\left(3\left(\omega_0 t + \varphi\right)\right) + 3\cos\left(\omega_0 t + \varphi\right)\right)\right). \tag{3.25}$$

Here we have used the relationship

$$(\cos\theta)^3 = \frac{1}{4}\cos 3\theta + \frac{3}{4}\cos\theta. \tag{3.26}$$

In the right-hand side of equation (3.25) we have two terms. The first is proportional to $\cos\left(3\left(\omega_0 t + \varphi\right)\right)$, and describes the generation of the third harmonic of the frequency ω_0. The second term, the so-called secular term, formally leads to the resonant growth with time of the first order solution: $x^{(1)} \sim t$. This conclusion is in contradiction with what we observed in the dependence of the potential (3.20) on x. We should expect to obtain dependence of the frequency of oscillations on their amplitude instead of the linear growth of x with time. For the frequency we should have

$$\omega = \omega_0 + \delta\omega, \tag{3.27}$$

where ω_0 is greater than $\delta\omega$. That means that ω_0 corresponds to fast oscillations, meanwhile $\delta\omega$ corresponds to relatively slow variations of the amplitude and phase.

Since we know that there is a correspondence between frequency and time derivative $d/dt \div -i\omega$ we can say that

$$-i\omega_0 \longleftrightarrow \frac{d}{dt_0}, \quad \text{and} \quad -i\delta\omega \longleftrightarrow \frac{d}{dt_1} \propto \varepsilon. \tag{3.28}$$

This observation provides us basis for the version of the perturbation theory called the multiple time expansion.

Multiple Time Expansion

We shall look for the solution of the Duffing equation (3.19) in the form

$$x = x^{(0)} + \varepsilon x^{(1)} + \varepsilon^2 x^{(2)} + \ldots, \tag{3.29}$$

where the function x depends on multiple times

$$x = x\left(t_0, t_1, t_2, \ldots\right). \tag{3.30}$$

Here t_0 is a fast time, t_1 is a slow time, t_2 is slower time, and so on.
    Expansion of the time derivative gives

$$\frac{d}{dt} = \frac{\partial}{\partial t_0} \frac{dt_0}{dt} + \frac{\partial}{\partial t_1} \frac{dt_1}{dt} + \frac{\partial}{\partial t_2} \frac{dt_2}{dt} + \ldots \tag{3.31}$$

We choose the time derivatives to obey the following ordering

$$\frac{dt_0}{dt} = 1, \quad \frac{dt_1}{dt} = \varepsilon, \quad \frac{dt_2}{dt} = \varepsilon^2, \ldots \tag{3.32}$$

Then

$$\frac{d}{dt} = \frac{\partial}{\partial t_0} + \varepsilon \frac{\partial}{\partial t_1} + \varepsilon^2 \frac{\partial}{\partial t_2} + \ldots, \tag{3.33}$$

$$\frac{d^2}{dt^2} = \frac{\partial^2}{\partial t_0^2} + 2\varepsilon \frac{\partial^2}{\partial t_0 \partial t_1} + \ldots \tag{3.34}$$

Substituting (3.29), (3.34) into equation (3.19) we obtain to zeroth order in ε equation

$$\frac{\partial^2 x^{(0)}}{\partial t_0^2} + \omega_0^2 x^{(0)} = 0 \tag{3.35}$$

with solution

$$x^{(0)} = a\left(t_1, t_2, \ldots\right) \cos\left(\omega_0 t_0 + \varphi\left(t_1, t_2, \ldots\right)\right); \tag{3.36}$$

to first order in ε equation

$$\frac{\partial^2 x^{(1)}}{\partial t_0^2} + \omega_0^2 x^{(1)} = -2 \frac{\partial^2 x^{(0)}}{\partial t_0 \partial t_1} - \left(x^{(0)}\right)^3 = -\frac{a^3}{4} \cos\left(3\left(\omega_0 t_0 + \varphi\right)\right)$$

$$-\frac{3a^3}{4}\cos\left(\omega_0 t_0 + \varphi\right) + 2\omega_0 \frac{\partial a}{\partial t_1} \sin\left(\omega_0 t_0 + \varphi\right)$$

$$+2\omega_0 a \frac{\partial \varphi}{\partial t_1} \cos\left(\omega_0 t_0 + \varphi\right)$$

$$= -\frac{a^3}{4}\cos\left(3\left(\omega_0 t_0 + \varphi\right)\right) + 2\omega_0 \frac{\partial a}{\partial t_1} \sin\left(\omega_0 t_0 + \varphi\right)$$

$$-\left(\frac{3a^2}{4} - 2\omega_0 \frac{\partial \varphi}{\partial t_1}\right) a \cos\left(\omega_0 t_0 + \varphi\right). \tag{3.37}$$

The secular terms in the right-hand side of the equation, which are the terms proportional to $\cos\left(\omega_0 t_0 + \varphi\right)$ and $\sin\left(\omega_0 t_0 + \varphi\right)$, respectively, must be canceled. This gives

$$2\omega_0 \frac{\partial a}{\partial t_1} = 0, \tag{3.38}$$

$$\frac{\partial \varphi}{\partial t_1} = \frac{3a^2}{8\omega_0}. \tag{3.39}$$

From these equations we find

$$a = \text{const}, \qquad \varphi = \frac{3a^2}{8\omega_0} t_1 \equiv \frac{3a^2 \beta}{8\omega_0} t, \tag{3.40}$$

and

$$x^{(0)} = a_0 \cos\left(\left(\omega_0 + \frac{3\beta}{8\omega_0} a^2\right) t_0 + \varphi_1\right). \tag{3.41}$$

We see that the frequency changes due to nonlinear effects with $\delta\omega$ proportional to the second power of the oscillation amplitude:

$$\omega = \omega_0 + \frac{3\beta}{8\omega_0} a^2 + \dots. \tag{3.42}$$

The solution to first order in ε is

$$x^{(1)} = b \cos\left(\omega_0 t_0 + \psi\right) + \frac{a_0^3}{32\omega_0^2} \cos\left(3\left(\omega_0 t_0 + \varphi\right)\right). \tag{3.43}$$

The last term in the right-hand side describes oscillations with the third harmonic of the frequency ω_0 which appear due to the nonlinear effects.

Asymptotic Expansion

We know well the Taylor series when in the vicinity of the point x_0 a function $f(x)$ can be represented in the form of the series in $(x - x_0)$:

$$f(x) = \sum_{n=0}^{\infty} \frac{f^{(n)}(x_0)}{n!} (x - x_0)^n. \tag{3.44}$$

The series $f_N(x) = \sum\limits_{n=0}^{N} \frac{f^{(n)}(x_0)}{n!}(x-x_0)^n$ converge for some R if $|x-x_0| < R$ when $N \to \infty$.

The Multiple-Time-Scale-Expansion provides an asymptotic expansion of a function $f(x)$. To perform an asymptotic expansion one needs to choose functions $\delta_n(\varepsilon)$ which are assumed to be small:

$$\delta_n(\varepsilon) = o(\delta_{n-1}(\varepsilon)) \qquad \text{as} \qquad \varepsilon \to 0. \qquad (3.45)$$

The function $\delta_n(\varepsilon)$ may be as it reads

$$\varepsilon^n, \quad \varepsilon^{n/2}, \quad (\varepsilon \ln \varepsilon)^n, \quad (\sin \varepsilon)^n, \quad \dots . \qquad (3.46)$$

A series $\sum\limits_{n=0}^{\infty} a_n \delta_n(\varepsilon)$ is said to be an asymptotic expansion of a function $f(\varepsilon)$ if

$$f(\varepsilon) - \sum_{n=0}^{N} a_n \delta_n(\varepsilon) = o(\delta_N(\varepsilon)) \qquad \text{as} \qquad \varepsilon \to 0 \qquad (3.47)$$

for every $N = 1, 2, \dots$. The series itself may be either convergent or divergent.

Example. The Bessel function $J_0(x)$ can be expanded into the Taylor series

$$J_0(x) = 1 - \frac{x^2}{2^2} + \frac{x^4}{2^2 4^2} - \frac{x^6}{2^2 4^2 6^2} + \dots . \qquad (3.48)$$

It converges for all x. Also the Bessel function $J_0(x)$ can be represented in the form of asymptotic expansion

$$J_0(x) \approx \left(\frac{2}{\pi x}\right)^{1/2} \left(u(x) \cos\left(x - \frac{\pi}{4}\right) + v(x) \sin\left(x - \frac{\pi}{4}\right) \right) \qquad (3.49)$$

for $x \to \infty$. Here the functions $u(x)$ and $v(x)$ are

$$u(x) = 1 - \frac{1^2 3^2}{2^2 4^2 2! x^2} + \dots, \qquad v(x) = \frac{1}{2 \cdot 4x} - \frac{1^2 3^2 5^2}{2^3 4^3 3! x^3} + \dots . \qquad (3.50)$$

The expansion (3.49) diverges when $x \to \infty$, however while n increases both the functions $u(x)$ and $v(x)$ decrease. We need to truncate the expansion.

Let us consider a function $f(x)$ defined by the integral

$$f(x) = \int_0^\infty \frac{\exp(-t)}{1 + xt} dt. \qquad (3.51)$$

Integrating by parts repeatedly we obtain

$$f(x) = 1 - x \int_0^\infty \frac{\exp(-t)}{(1 + xt)^2} dt = 1 - x + 2x^2 \int_0^\infty \frac{\exp(-t)}{(1 + xt)^3} dt = \dots$$

$$= \sum_{n=0}^{N}(-1)^n n! x^n + (-1)^{N+1}(N+1)! x^{N+1} \int_0^\infty \frac{\exp(-t)}{(1+xt)^{N+2}} dt. \qquad (3.52)$$

The last integral is $O(1)$ as $x \to 0$. This is Euler's asymptotic expansion of the integral (3.51).

●

3.3   Nonlinear Resonance

We generalize the Duffing equation (3.19) to describe the nonlinear resonance

$$\ddot{x} + 2\nu\dot{x} + \omega_0^2 x + \beta x^3 = f \cos \Omega t. \qquad (3.53)$$

Here ν, β and f are assumed to be small and to be of the same order as $\varepsilon \ll 1$.
    Multiple time scale expansion gives to zeroth order in ε

$$\frac{\partial^2 x^{(0)}}{\partial t_0^2} + \omega_0^2 x^{(0)} = 0, \qquad (3.54)$$

$$x^{(0)} = a \cos \theta \qquad \text{with} \qquad \theta = \omega_0 t_0 + \varphi. \qquad (3.55)$$

The first order in ε gives

$$\frac{\partial^2 x^{(1)}}{\partial t_0^2} + \omega_0^2 x^{(1)} = f \cos \Omega t + \left(2\omega_0 \frac{\partial a}{\partial t_1} + 2\nu\omega_0 a \right) \sin \theta +$$

$$\left(2\omega_0 a \frac{\partial \varphi}{\partial t_1} - \frac{3\beta a^3}{4} \right) \cos \theta + \frac{\beta a^3}{4} \cos 3\theta. \qquad (3.56)$$

We introduce the value δ for the frequency difference $\Omega - \omega_0 = \varepsilon\delta$ with $\varepsilon\delta$ being of the order of ε. Then we can write

$$\cos \Omega t = \cos(\theta + \delta t_1 - \varphi) =$$

$$\cos \theta \cos(\delta t_1 - \varphi) - \sin \theta \sin(\delta t_1 - \varphi). \qquad (3.57)$$

Substituting this expression into equation (3.35) we obtain

$$\frac{\partial^2 x^{(1)}}{\partial t_0^2} + \omega_0^2 x^{(1)} = + \left(2\omega_0 \frac{\partial a}{\partial t_1} + 2\nu\omega_0 a - f \sin(\delta t_1 - \varphi) \right) \sin \theta +$$

$$\left(2\omega_0 a \frac{\partial \varphi}{\partial t_1} - \frac{3\beta a^3}{4} + f \cos(\delta t_1 - \varphi) \right) \cos \theta + \frac{\beta a^3}{4} \cos 3\theta. \qquad (3.58)$$

First two terms in the right-hand side are secular. They must be put to equal zero. That gives equations for a and φ :

$$2\omega_0 \frac{\partial a}{\partial t_1} + 2\nu\omega_0 a = f \sin(\delta t_1 - \varphi), \qquad (3.59)$$

$$2\omega_0 a \frac{\partial \varphi}{\partial t_1} - \frac{3\beta a^3}{4} = -f \cos (\delta t_1 - \varphi) . \tag{3.60}$$

We introduce the function

$$\psi = \delta t_1 - \varphi; \tag{3.61}$$

as the result the system of equations (3.59–3.60) takes the form

$$\frac{\partial a}{\partial t_1} = -2\nu a + \frac{f}{2\omega_0} \sin \psi, \tag{3.62}$$

$$\frac{\partial \psi}{\partial t_1} = \frac{3\beta}{8\omega_0} a^2 - \frac{f}{2a\omega_0} \cos \psi. \tag{3.63}$$

These equations in the limit $\beta = 0$ describe the resonance in the linear system when the dissipation effects (the term $2\nu\dot{x}$ is added) are incorporated into equation (3.3). We seek the solution with $\psi = \pi/2$. Then we obtain from (3.62) and (3.63)

$$\frac{\partial \psi}{\partial t_1} = 0, \qquad \frac{\partial a}{\partial t_1} + 2\nu a = \frac{f}{2\omega_0} \tag{3.64}$$

with a solution

$$a(t) = \frac{f}{2\nu\omega_0} (1 - \exp(-\nu t)). \tag{3.65}$$

When $\nu t \ll 1$ we have

$$a(t) \approx \frac{f}{2\omega_0} t. \tag{3.66}$$

This expression describes a linear growth of the oscillation amplitude with time. In the limit $\nu t \gg 1$ there is a saturation of the resonance: the amplitude reaches its maximum value

$$a \approx \frac{f}{2\nu\omega_0} \tag{3.67}$$

in accordance with what is shown in Fig. 3.1.

Now we consider the case $\beta \neq 0$. The steady state solution corresponds to $\partial/\partial t_1 = 0$. It is the saturated regime of oscillations in the nonlinear resonance. Using the relationship $\sin^2 \psi + \cos^2 \psi = 1$ we obtain from equations (3.62), (3.63) the relationship

$$\frac{f^2}{4a^2\omega_0^2} = \nu^2 + \left(\delta - \frac{3\beta}{8\omega_0} a^2 \right)^2 , \tag{3.68}$$

which gives the dependence of the amplitude a on f, Ω, ω_0, ν, and β. The dependence of a on δ is shown in Fig. 3.3.

Here we see the nonlinear shift of the frequency that leads to detuning between the driven force and oscillations while the amplitude grows. Rewriting equation (3.68) we obtain

$$\delta = \frac{3\beta}{8\omega_0} a^2 \pm \left(\frac{f^2}{4a^2\omega_0^2} - \nu^2 \right)^{1/2} \tag{3.69}$$

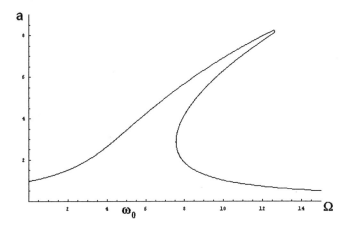

Fig. 3.3. Nonlinear resonance.

an implicit expression for a as a function of δ.

Exercise. Find the nonlinear shift of the frequency for nonlinear oscillations described by equation

$$\ddot{x} + \omega_0^2 x + \alpha x^2 = 0. \tag{3.70}$$

$\alpha \ll 1$.

$\diamond$

Exercise. Investigate the nonlinear resonance for equation

$$\ddot{x} + 2\nu\dot{x} + \omega_0^2 x + \alpha x^2 = f \cos \Omega t. \tag{3.71}$$

$\diamond$

3.3.1 Limit Circle. The Van der Pol Equation.

The Van der Pol equation is

$$\ddot{x} + \omega_0^2 x - \varepsilon\omega_0 \left(1 - \beta\dot{x}^2\right)\dot{x} = 0. \tag{3.72}$$

Depending on the sign of the parameter β it describes two different types of the oscillation behavior as it is shown in Fig. 3.4 a and 3.4 b.

If $\beta > 0$, the point $x = 0, \dot{x} = 0$ is unstable. A trajectory in the phase plane $(x, \dot{x})$ tends to the limit circle (see Fig. 3.4 a). The limit circle in this case is an attractor, while the point $x = 0, \dot{x} = 0$ is a repeller. It attracts trajectories. When $\beta < 0$ the point $x = 0, \dot{x} = 0$ is stable. A trajectory in the phase plane $(x, \dot{x})$ does not tend to the limit circle (see Fig. 3.4 b). It leaves the neighborhood of the limit circle. In this case it is a repeller, while the point $x = 0, \dot{x} = 0$ is an attractor.

To solve the Van der Pol equation we use the multiple time expansion theory. We suppose ε and β to be small and of the same order.

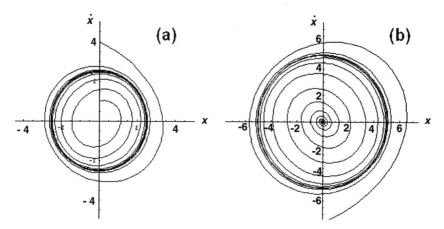

Fig. 3.4. Stable (a) and unstable (b) limit circles.

Multiple time scale expansion gives to zeroth order in ε

$$\frac{\partial^2 x^{(0)}}{\partial t_0^2} + \omega_0^2 x^{(0)} = 0, \tag{3.73}$$

$$x^{(0)} = a \cos \theta \quad \text{with} \quad \theta = \omega_0 t_0 + \varphi. \tag{3.74}$$

The first order in ε gives

$$\frac{\partial^2 x^{(1)}}{\partial t_0^2} + \omega_0^2 x^{(1)} = \left(2\omega_0 \frac{\partial a}{\partial t_1} - \omega_0^2 a + \frac{\beta \omega_0^3}{4} a^3 \right) \sin \theta +$$

$$2\omega_0 \frac{\partial \varphi}{\partial t_1} a \cos \theta + \frac{\beta \omega_0^2}{4} a^3 \cos 3\theta. \tag{3.75}$$

Here we have used the formula

$$\cos^2 \theta \sin \theta = \frac{1}{4} \sin \theta + \frac{1}{4} \sin 3\theta. \tag{3.76}$$

Cancelling the secular terms we find

$$2\omega_0 \frac{\partial a}{\partial t_1} - \omega_0^2 a + \frac{\beta \omega_0^3}{4} a^3 = 0, \tag{3.77}$$

$$2\omega_0 a \frac{\partial \psi}{\partial t_1} = 0. \tag{3.78}$$

Solution to equation (3.78) gives $\psi = \text{const}$,. Let it be equal to zero. For the amplitude a from equation (3.79) we find the expression

$$\int_{a_0}^{a} \frac{da}{a \left(1 - \beta a^2 / 4 \right)} = \frac{\omega_0}{2} t_1, \tag{3.79}$$

that gives

$$a(t) = \frac{a_0}{\left(\left(1 - \frac{\beta a_0^2}{4}\right)\exp(-\varepsilon\omega_0 t) + \frac{\beta a_0^2}{4}\right)^{1/2}}. \qquad (3.80)$$

When $t \to \infty$ the amplitude tends to $2/\beta^{1/2}$. That means that the amplitude of oscillations on the limit circle is equal to $2/\beta^{1/2}$.

3.3.2 Linear Oscillator with a Slowly Varying Frequency

We consider equation

$$\ddot{x} + \omega_0^2(\varepsilon t)x = 0, \qquad (3.81)$$

where frequency ω_0 depends on time and the parameter ε is supposed to be small. To solve the equation we use the multiple time expansion method in the form when $x(t)$ is represented as

$$x = x(\theta, t_1, t_2, \ldots) = x^{(0)} + \varepsilon x^{(1)} + \varepsilon^2 x^{(2)} + \ldots, \qquad (3.82)$$

with

$$\frac{d}{dt} = \frac{\partial}{\partial\theta}\frac{\partial\theta}{\partial t} + \frac{\partial}{\partial t_1}\frac{\partial t_1}{\partial t} + \ldots. \qquad (3.83)$$

The fast-time-variable θ is chosen as

$$\theta = \int^t \omega_0(\varepsilon t)\,dt = \frac{1}{\varepsilon}\int^{\varepsilon t} \omega_0(\xi)\,d\xi \equiv \theta(t_1) = \frac{g(t_1)}{\varepsilon} \qquad (3.84)$$

with $t_1 = \varepsilon t$.

The expansion of the time derivative gives

$$\frac{d^2}{dt^2} = \omega_0^2(t_1)\frac{\partial^2}{\partial\theta^2} + \varepsilon\left(2\omega_0(t_1)\frac{\partial^2}{\partial\theta\partial t_1} + \frac{\partial\omega_0}{\partial t_1}\frac{\partial}{\partial\theta}\right) + \ldots. \qquad (3.85)$$

Substituting expression (3.85) into equation (3.81) we obtain to zeroth order in ε

$$\omega_0^2(t_1)\left(\frac{\partial^2 x^{(0)}}{\partial\theta^2} + x^{(0)}\right) = 0, \qquad (3.86)$$

$$x^{(0)} = a\cos(\theta + \varphi). \qquad (3.87)$$

The first order in ε gives

$$\omega_0^2(t_1)\left(\frac{\partial^2 x^{(1)}}{\partial\theta^2} + x^{(1)}\right) = -\left(2\omega_0(t_1)\frac{\partial^2 x^{(0)}}{\partial\theta\partial t_1} + \frac{\partial\omega_0}{\partial t_1}\frac{\partial x^{(0)}}{\partial\theta}\right) =$$

$$\left(2\omega_0(t_1)\frac{\partial a}{\partial t_1} + \frac{\partial\omega_0}{\partial t_1}a\right)\sin(\theta + \varphi) + 2\omega_0(t_1)a\frac{\partial\varphi}{\partial t_1}\cos(\theta + \varphi). \qquad (3.88)$$

From this equation cancelling the secular terms we obtain equations

$$2\omega_0\left(t_1\right)\frac{\partial a}{\partial t_1} + \frac{\partial \omega_0}{\partial t_1}a = 0, \tag{3.89}$$

and

$$2\omega_0\left(t_1\right)a\frac{\partial \varphi}{\partial t_1} = 0 \tag{3.90}$$

with the solutions

$$a(t) = a(0)\left(\frac{\omega_0\left(0\right)}{\omega_0\left(t_1\right)}\right)^{1/2}, \qquad \varphi = 0. \tag{3.91}$$

Finally we have

$$x(t) = a(0)\left(\frac{\omega_0\left(0\right)}{\omega_0\left(t_1\right)}\right)^{1/2}\cos\left(\frac{1}{\varepsilon}\int^{\varepsilon t}\omega_0\left(\xi\right)d\xi\right). \tag{3.92}$$

Corollary. This result can also be obtained within the WKB approach. There is an overlapping of the ranges of validity of both the methods.
$\star$

3.3.3   WKB Approximation

The solutions that are called WKB solutions were used by *Wentzel (1928)*, *Kramers (1926)*, and *Brillouin (1926)* in quantum mechanics. Let us consider the WKB approximation in regard with equation (3.81). In the stationary system there are two solutions which present oscillatory motions:

$$x(t) = a\exp(i\theta(t)), \tag{3.93}$$

where the function $\theta(t)$, in our case $\theta(t) = \pm\omega_0 t$, is the generalized phase of oscillations. For complex $\omega_0^2(t)$ it is complex. We assume the amplitude, a, to be constant. Differentiating $x(t)$ with respect to time we obtain

$$\ddot{x} = a(i\ddot{\theta} - \dot{\theta}^2)\exp(i\theta(t)). \tag{3.94}$$

Substituting this into equation (3.81) we see that $\theta(t)$ must satisfy

$$\dot{\theta}^2 = \omega_0^2(t) + i\ddot{\theta}. \tag{3.95}$$

Instead of a linear differential equation (3.81) we have obtained a nonlinear differential equation which is more difficult to solve analytically. To find an approximate solution we use the fact that dependence of ω_0^2 on t has been supposed to be slow. In this case $\ddot{\theta}$ is small compared to both $\dot{\theta}^2$ and ω_0^2. Then

$$\dot{\theta} \approx \pm\omega_0 \qquad \text{and} \qquad \ddot{\theta} \approx \pm\varepsilon\dot{\omega}_0. \tag{3.96}$$

Substituting these approximate relationships into equation (3.95) we find a second approximation:

$$\dot{\theta} \approx \pm \left(\omega_0^2 \pm i\varepsilon\dot{\omega}_0\right)^{1/2} \approx \pm\omega_0 + \frac{i\varepsilon}{2\omega_0}\dot{\omega}_0. \tag{3.97}$$

Integrating the right-hand side yields

$$\theta \approx \pm \int^t \omega_0(\varepsilon t)dt + i\ln(\omega_0^{1/2}(\varepsilon t)), \tag{3.98}$$

and

$$x(t) \approx a\omega_0^{-1/2}(\varepsilon t)\exp\left(\pm i\int^t \omega_0(\varepsilon t)dt\right), \tag{3.99}$$

which for real ω_0 is the same as given by (3.92).

Exercise. Study nonstationary resonance described by equation

$$\ddot{x} + \omega_0^2(\varepsilon t)x = f\cos\Omega t \tag{3.100}$$

assuming that at the time $t = t_*$ $\omega_0(\varepsilon t_*) = \Omega$ with $t_* \gg \omega_0^{-1}$. A similar approach can be used to find the solution to equations with a small parameter in front of high-order-derivative:

$$\tilde{\varepsilon}\ddot{x} + \omega_0^2(t)x = 0. \tag{3.101}$$

The change of variables $\tau = (\tilde{\varepsilon})^{1/2}t$ transforms this equation to (3.81).

◇

Exercise. Using of the multiple time expansion method find the solution to equation

$$\varepsilon^\alpha\ddot{x} + \omega_0^2(\varepsilon^\beta t)x = 0 \tag{3.102}$$

for $2\alpha \neq \beta$ and $\varepsilon \ll 1$.

◇

The energy of oscillations equals the sum of kinetic and potential energy. It is

$$E(t) = \frac{\dot{x}^2 + \omega_0^2 x^2}{2} = \overline{\omega^2 x^2} \equiv a^2(0)\omega_0(0)\omega_0(\varepsilon t) = E(0). \tag{3.103}$$

That means that the quantity

$$J = \frac{E(t)}{\omega_0(t)} = \text{const} \tag{3.104}$$

is conserved. It is an *adiabatic invariant*. When the frequency of oscillations depends on time the energy does not conserve any more. However the value of J is constant if the change of the frequency ω_0 is slow.

3.3.4   Parametric Resonance

Let us assume the time dependence of the frequency of the form

$$w(t) = w\,(1 + \varepsilon \cos\,(2w_0 t))^{1/2}. \tag{3.105}$$

Parametric resonance is described by the equation

$$\ddot{x} + w^2\,(1 + \varepsilon \cos\,(2w_0 t))\,x = 0. \tag{3.106}$$

We consider the case when the value of w is close to that of w_0 :

$$w = w_0 + \varepsilon w^{(1)} + \dots. \tag{3.107}$$

The multiple time scale expansion gives

$$x = x^{(0)} + \varepsilon x^{(1)} + \varepsilon^2 x^{(2)} + \dots, \tag{3.108}$$

$$\frac{d}{dt} = \frac{\partial}{\partial t_0} + \varepsilon \frac{\partial}{\partial t_1} + \varepsilon^2 \frac{\partial}{\partial t_2} + \dots. \tag{3.109}$$

Substituting expressions (3.107), (3.108) and (3.109) into equation (3.106) we find to zeroth order in ε :

$$\frac{\partial^2 x^{(0)}}{\partial t_0^2} + w_0^2 x^{(0)} = 0 \tag{3.110}$$

with solution

$$x^{(0)} = a \cos w_0 t + b \sin w_0 t. \tag{3.111}$$

To first order in ε we obtain

$$\frac{\partial^2 x^{(1)}}{\partial t_0^2} + w_0^2 x^{(1)} = -2 \frac{\partial^2 x^{(0)}}{\partial t_0 \partial t_1} - 2w_0 w^{(1)} x^{(0)} - w_0^2 x^{(0)} \cos 2w_0 t_0. \tag{3.112}$$

Then we use the formula

$$x^{(0)} \cos 2w_0 t_0$$
$$= \frac{a}{2}(\cos 3w_0 t_0 + \cos w_0 t_0) + \frac{b}{2}(\sin 3w_0 t_0 - \sin w_0 t_0). \tag{3.113}$$

The secular terms in the right-hand side of equation (3.113) vanish when

$$-2\frac{\partial a}{\partial t_1} - 2w^{(1)} b + \frac{w_0}{2} b = 0, \tag{3.114}$$

$$2\frac{\partial b}{\partial t_1} + 2w^{(1)} a + \frac{w_0}{2} a = 0. \tag{3.115}$$

We look for the solution to equations (3.114)—(3.115) in the form $\sim \exp \gamma t$. For γ we have the dispersion equation

$$\det \begin{pmatrix} -2\gamma & -2w^{(1)} + w_0/2 \\ 2w^{(1)} + w_0/2 & 2\gamma \end{pmatrix} = 0, \tag{3.116}$$

or

$$\gamma^2 = \frac{\omega_0^2}{16} - \left(\omega^{(1)}\right)^2. \tag{3.117}$$

Parametric instability occurs when γ^2 is real, that is, when

$$\omega^{(1)} < \omega_0/4. \tag{3.118}$$

This expression can be rewritten as

$$\Delta\omega = \varepsilon\omega^{(1)} < \varepsilon\frac{\omega_0}{4} \ll \omega_0. \tag{3.119}$$

In this case we have

$$x^{(0)} = \left(a_0 \cosh(\gamma t) + b_0 \left(\frac{\omega_0 - 4\omega^{(1)}}{4\gamma}\right) \sinh(\gamma t)\right) \cos\omega_0 t$$

$$+ \left(b_0 \cosh(\gamma t) - a_0 \left(\frac{\omega_0 + 4\omega^{(1)}}{4\gamma}\right) \sinh(\gamma t)\right) \sin\omega_0 t, \tag{3.120}$$

where $a_0 = x(0)$ and $b_0 = \dot{x}(0)/\omega_0$.

Motion in the High Frequency Field.

Above we have analyzed several cases of the particle motion in the time dependent potential wells. In the first case (parametric resonance) the potential well changed periodically with the frequency about two times higher than the frequency of the particle motion in the time independent potential. We have seen that the amplitude of oscillation grows exponentially as it is given by expression (3.120). In the second case of slowly varying potential well the adiabatic invariants are conserved. Then if the external force is periodic with the frequency of the order of the free oscillation, the frequency amplitude of driven motion grows proportionally to time in linear system or saturates due to nonlinear change of the frequency. Now we consider the case when the particle is under the action of the periodic force, with the frequency much higher that the frequency of free oscillations.

The equation of motion has the form

$$\ddot{x} = -\frac{\partial U}{\partial x} + f(x)\cos(\Omega t), \tag{3.121}$$

where the dependence of both the high-frequency force, $f(x)$, and potential, $U(x)$, on the x−coordinate is assumed to have the scale-length much larger than the amplitude of oscillations with the frequency Ω. We seek for the solution in the form

$$x(t) = x^{(0)}(t) + x^{(1)}(t), \tag{3.122}$$

with $x^{(0)}(t)$ and $x^{(1)}(t)$ to be the low- and high-frequency functions. That is, $\langle x(t) \rangle = x^{(0)}(t)$ and $\langle x^{(1)}(t) \rangle = 0$. A notation $\langle ... \rangle$ stands for the time averaging on the high frequency: $\langle ... \rangle = \frac{\Omega}{2\pi} \int_0^{2\pi/\Omega} (...)dt$.

Substituting expression (3.122) into equation (3.121) and expanding in $x^{(1)}$, we obtain

$$\ddot{x}^{(0)} + \ddot{x}^{(1)} = -\left.\frac{\partial U}{\partial x}\right|_{x^{(0)}} + f(x^{(0)})\cos(\Omega t) + \left(\left.\frac{\partial f}{\partial x}\right|_{x^{(0)}}\cos(\Omega t) - \left.\frac{\partial^2 U}{\partial x^2}\right|_{x^{(0)}}\right)x^{(1)}.$$
$$(3.123)$$

Further we need to separate the low- and high-frequency parts. Averaging on the high-frequency time scale, we obtain

$$\ddot{x}^{(0)} = -\left.\frac{\partial U}{\partial x}\right|_{x^{(0)}} + \left\langle \left.\frac{\partial f}{\partial x}\right|_{x^{(0)}}\cos(\Omega t)x^{(1)}\right\rangle \tag{3.124}$$

for low-frequency motion. To find the equation for high-frequency motion we subtract equation (3.123) from equation (3.124). Neglecting the small term,

$$\left.\frac{\partial f}{\partial x}\right|_{x^{(0)}}\cos(\Omega t)x^{(1)} - \left\langle \left.\frac{\partial f}{\partial x}\right|_{x^{(0)}}\cos(\Omega t)x^{(1)}\right\rangle,$$

we find for high-frequency motion

$$\ddot{x}^{(1)} + \left.\frac{\partial^2 U}{\partial x^2}\right|_{x^{(0)}} x^{(1)} = f(x^{(0)})\cos(\Omega t). \tag{3.125}$$

Since the frequency of free oscillations, $\omega_0 = \left(\left.\frac{\partial^2 U}{\partial x^2}\right|_{x^{(0)}}\right)^{1/2}$, is assumed to be much less than Ω, we neglect the second term in the left hand side of equation (44). This yields

$$x^{(1)}(t) = -\frac{f(x^{(0)})}{\Omega^2}\cos(\Omega t). \tag{3.126}$$

We substitute $x^{(1)}(t)$ into last term in the right-hand side of equation (3.124) and obtain that averaged on high frequency motion obeys equation

$$\ddot{x}^{(0)} = -\frac{\partial U_{\text{eff}}}{\partial x} \tag{3.127}$$

with effective potential

$$U_{\text{eff}} = U + \frac{f^2}{2\Omega^4}.$$

The force $-\nabla(f^2/2\Omega^4)$ is called *the ponderomotive force* .

3.3.5   Motion of Systems with Two Degrees of Freedom

Let us consider free oscillations of a system of two coupled nonlinear oscillators having different frequencies, ω_1 and ω_2, damping rates, ν_1 and ν_2. We assume that the interaction between the oscillators can be described by quadratic nonlinear terms in the right-hand sides of equations

$$\ddot{x} + 2\nu_1\dot{x} + \omega_1^2 x = \alpha_1 xy, \tag{3.128}$$

$$\ddot{y} + 2\nu_2\dot{y} + \omega_2^2 y = \alpha_2 y^2. \tag{3.129}$$

We assume that damping and nonlinearity are weak being of the order of $\varepsilon \ll 1$.

To solve the system (3.128)-(3.129) we use a multiple time expansion. We seek for the solution with both the functions x and y to depend on multiple times

$$x = x(t_0, t_1, t_2, \ldots), \qquad y = y(t_0, t_1, t_2, \ldots). \tag{3.130}$$

Here t_0 is a fast time, t_1 is a slow time, t_2 is a slower time, and so on.

Expansion of the functions $x(t)$ and $y(t)$ and time derivative gives

$$x = x^{(0)} + x^{(1)} + \cdots, \qquad y = y^{(0)} + y^{(1)} + \cdots. \tag{3.131}$$

and

$$\frac{d}{dt} = \frac{\partial}{\partial t_0} + \varepsilon \frac{\partial}{\partial t_1} + \varepsilon^2 \frac{\partial}{\partial t_2} + \cdots, \tag{3.132}$$

$$\frac{d^2}{dt^2} = \frac{\partial^2}{\partial t_0^2} + 2\varepsilon \frac{\partial^2}{\partial t_0 \partial t_1} + \cdots. \tag{3.133}$$

Substituting these expressions into equations (3.128)-(3.129) we obtain to zeroth order in ε equations

$$\frac{\partial^2 x^{(0)}}{\partial t_0^2} + \omega_1^2 x^{(0)} = 0, \tag{3.134}$$

$$\frac{\partial^2 y^{(0)}}{\partial t_0^2} + \omega_2^2 y^{(0)} = 0, \tag{3.135}$$

with solution

$$x^{(0)} = a(t_1, \ldots) \exp(i\omega_1 t_0) + c.c. \quad \text{and} \quad y^{(0)} = b(t_1, \ldots) \exp(i\omega_2 t_0) + c.c., \tag{3.136}$$

where "c.c." stands for complex conjugate terms. To first order in ε we obtain equations

$$\frac{\partial^2 x^{(1)}}{\partial t_0^2} + \omega_1^2 x^{(1)} = -2\frac{\partial^2 x^{(0)}}{\partial t_0 \partial t_1} - 2\bar{\nu}_1 \frac{\partial x^{(0)}}{\partial t_0} + \bar{\alpha}_1 x^{(0)} y^{(0)} =$$

$$-2i\omega_1 \left(\frac{\partial a}{\partial t_1} + \bar{\nu}_1 a \right) \exp(i\omega_1 t_0)$$

$$+\bar{\alpha}_1 \left(ab \exp(i(\omega_1 + \omega_2)t_0) + a^* b \exp(i(\omega_2 - \omega_1)t_0) \right) + c.c., \tag{3.137}$$

$$\frac{\partial^2 y^{(1)}}{\partial t_0^2} + \omega_2^2 y^{(1)} = -2\frac{\partial^2 y^{(0)}}{\partial t_0 \partial t_1} - 2\bar{\nu}_2 \frac{\partial y^{(0)}}{\partial t_0} + \bar{\alpha}_2 (x^{(0)})^2 =$$

$$-2i\omega_2 \left(\frac{\partial b}{\partial t_1} + \bar{\nu}_2 b \right) \exp(i\omega_2 t_0)$$

$$+\bar{\alpha}_2 \left(a^2 \exp(i2\omega_1 t_0) + ab^* \right) + c.c.. \tag{3.138}$$

Here $\bar{\nu}_1 = \nu_1/\varepsilon$, $\bar{\alpha}_1 = \alpha_1/\varepsilon$, $\bar{\nu}_2 = \nu_2/\varepsilon$ and $\bar{\alpha}_2 = \alpha_2/\varepsilon$.

Now we must distinguish two cases depending on whether we have nonresonant, $w_2 \neq 2w_1$, or resonant case with $w_2 \approx 2w_1$.

Nonresonant case

Since $w_2 \neq 2w_1$ secular terms are the first terms in the right-hand sides of equations (3.137) and (3.138). Solvability conditions for these equations are

$$\left(\frac{\partial a}{\partial t_1} + \bar{\nu}_1 a \right) = 0, \tag{3.139}$$

$$\left(\frac{\partial b}{\partial t_1} + \bar{\nu}_2 b \right) = 0. \tag{3.140}$$

We see that

$$a = a_0 \exp(-\bar{\nu}_1 t_1) \qquad \text{and} \qquad b = b_0 \exp(-\bar{\nu}_2 t_1). \tag{3.141}$$

Solutions of equations (3.137) and (3.138) can be written as

$$x(t) = (a_0 \exp(i w_1 t) + c.c.) \exp(-\nu_1 t) + O(\varepsilon^2), \tag{3.142}$$

$$y(t) = (b_0 \exp(i w_2 t) + c.c.) \exp(-\nu_2 t) + O(\varepsilon^2). \tag{3.143}$$

We see that both $x(t)$ and $y(t)$ tend to zero as $t \to \infty$.

Internal resonance

In the case of internal resonance we have $w_2 = 2w_1 + \delta w$, where the detuning, δw, is assumed to be small: $\delta w = \varepsilon w^{(1)}$. We express $2w_1 t_0$ and $(w_2 - w_1) t_0$ in the right-hand sides of equations (3.137) and (3.138) as

$$2w_1 t_0 = w_2 t_0 - \varepsilon w^{(1)} t_0 = w_2 t_0 - w^{(1)} t_1, \tag{3.144}$$

$$(w_2 - w_1) t_0 = w_1 t_0 + \varepsilon w^{(1)} t_0 = w_1 t_0 + w^{(1)} t_1. \tag{3.145}$$

Solvability conditions for equations (3.137) and (3.138) are

$$-2w_1 \left(\frac{\partial a}{\partial t_1} + \bar{\nu}_1 a \right) + \bar{\alpha}_1 a^* b \exp(i w^{(1)} t_1) = 0, \tag{3.146}$$

$$-2w_2 \left(\frac{\partial b}{\partial t_1} + \bar{\nu}_2 b \right) + \bar{\alpha}_2 a^2 \exp(-i w^{(1)} t_1) = 0. \tag{3.147}$$

It is convenient to introduce polar notations. We put

$$a = |a| \exp(i\theta) \qquad \text{and} \qquad b = |b| \exp(i\varphi) \tag{3.148}$$

and rewrite equations (3.146) and (3.147) as

$$\frac{\partial |a|}{\partial t_1} = -\bar{\nu}_1 |a| + \frac{\bar{\alpha}_1}{2w_1} |a||b| \sin \psi, \tag{3.149}$$

$$\frac{\partial |b|}{\partial t_1} = -\bar{\nu}_2 |b| - \frac{\bar{\alpha}_2}{2\omega_2} |a|^2 \sin \psi, \tag{3.150}$$

$$|a| \frac{\partial \theta}{\partial t_1} = -\frac{\bar{\alpha}_1}{2\omega_1} |a||b| \cos \psi, \tag{3.151}$$

$$|b| \frac{\partial \varphi}{\partial t_1} = -\frac{\bar{\alpha}_2}{2\omega_2} |a|^2 \cos \psi, \tag{3.152}$$

where the new phase is

$$\psi = \varphi - 2\theta + \omega^{(1)} t_1. \tag{3.153}$$

Eliminating θ and φ from (3.151) and (3.152) through (3.153) yields

$$|b| \frac{\partial \psi}{\partial t_1} = \omega^{(1)} |b| + \left(\frac{\bar{\alpha}_1}{\omega_1} |b|^2 - \frac{\bar{\alpha}_2}{2\omega_2} |a|^2 \right) \cos \psi. \tag{3.154}$$

Steady state response corresponds to $\partial |a|/\partial t_1 = \partial |b|/\partial t_1 = \partial \psi/\partial t_1 = 0$. Thus we have

$$-\bar{\nu}_1 |a| + \frac{\bar{\alpha}_1}{2\omega_1} |a||b| \sin \psi = 0, \tag{3.155}$$

$$-\bar{\nu}_2 |b| - \frac{\bar{\alpha}_2}{2\omega_2} |a|^2 \sin \psi = 0, \tag{3.156}$$

$$\omega^{(1)} |b| + \left(\frac{\bar{\alpha}_1}{\omega_1} |b|^2 - \frac{\bar{\alpha}_2}{2\omega_2} |a|^2 \right) \cos \psi = 0, \tag{3.157}$$

which give

$$|a|^2 \frac{\nu_1 \omega_1}{\alpha_1} + |b|^2 \frac{\nu_2 \omega_2}{\alpha_2} = 0. \tag{3.158}$$

We see that for steady state response ν_1 and ν_2 must have opposite signs. In this case ,

$$|a| = \left(-\frac{4\omega_1^3 \nu_1 \nu_2}{\omega_2 \alpha_1 \alpha_2} \left(1 + \frac{\omega_2^4 \left(\omega^{(1)} \right)^2}{(2\nu_1 \omega_2^2 + \nu_2 \omega_1^2)^2} \right) \right)^{1/2}, \tag{3.159}$$

$$|b| = \frac{2\omega_1 \nu_1}{\alpha_1} \left(1 + \frac{\omega_2^4 \left(\omega^{(1)} \right)^2}{(2\nu_1 \omega_2^2 + \nu_2 \omega_1^2)^2} \right)^{1/2}. \tag{3.160}$$

When both ν_1 and ν_2 vanish the problem can be solved in terms of the elliptic functions.

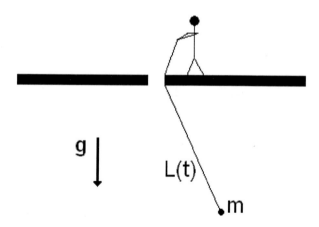

Fig. 3.5. Pendulum with varying length $L(t)$.

3.4 Adiabatic Invariants

Adiabatic invariants. Accuracy of conservation of adiabatic invariants. Action-angle canonical variables.

***

3.4.1 Adiabatic Motion

Above we have solved the equation which describes oscillations with slowly varying frequency and found that there is conserved quantity called the adiabatic invariant, which in this case is given by formula

$$J = \frac{E(t)}{\omega_0(t)}. \tag{3.161}$$

A pendulum with the threat which length $L(t)$ varying slowly with time provides the standard example of such a system with conserved adiabatic invariant, $J = E(t)/\omega_0(t)$. Here the frequency of small amplitude oscillation of the pendulum equals $\omega_0(t) = (g/L(t))^{1/2}$. Such a pendulum is shown in Fig. 3.5.

While the pendulum length, $L(t)$, is changed, both the pendulum motion energy and frequency change. However, the action function $J(q, p, L(t))$, which by definition is $(2\pi)^{-1}$ times the instantaneous area of the Hamiltonian $\mathcal{H}(q, p, L(t)) = \mathcal{H}$ contour in the phase plane (p, q), is adiabatically conserved.

Example. *Transverse Adiabatic Invariant of a Charged Particle in a Magnetic Field.* When the magnetic field dependence on the coordinates, $\mathbf{B}(\varepsilon \mathbf{r})$, is assumed to be smooth, the motion of a charged particle can be presented as a motion along magnetic field lines, $\overline{r}_\parallel(t)$, slow drift perpendicularly to the

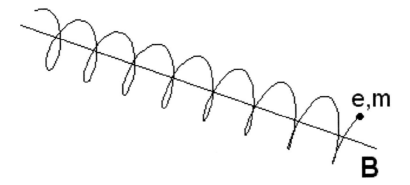

Fig. 3.6. The trajectory of a charged particle in the magnetic field.

direction of the magnetic field, $\overline{\mathbf{r}_\perp}(t)$, and fast rotation, $\widetilde{\mathbf{r}_\perp}(t)$, with the Larmor frequency $\omega_B = eB/mc$: $\widetilde{r_\perp}(t) = r_B \cos\omega_B t$, and the Larmor radius $r_B = mv_\perp c/eB$ (see Fig. 3.6).

Larmor rotation of a charged particle corresponds to the motion across the magnetic field. That is why the kinetic energy of transverse motion, $E_\perp$, can be written as

$$E_\perp = \frac{m\dot{r}^2}{2} = \frac{mr_B^2\omega_B^2}{2} = \frac{mv_\perp^2}{2}. \tag{3.162}$$

The transverse adiabatic invariant $J_\perp$ can be expressed as

$$J_\perp = \frac{E_\perp}{\omega_B} \propto \frac{E_\perp}{B}. \tag{3.163}$$

A more common notation for this adiabatic invariant, $\mu = mv_\perp^2/2B$, called *magnetic momentum*, is widely used.

•

Example.   *Longitudinal adiabatic invariant.* When a charged particle moves in the magnetic field with the magnitude of the magnetic field growing along the direction of the motion, the particle can reflect and may move in the opposite direction. To explain this phenomenon we need to invoke the conservation of particle energy and the transverse adiabatic invariant, μ. From the energy conservation we have

$$E = \frac{mv_\parallel^2}{2} + \frac{mv_\perp^2}{2} = \text{const.} \tag{3.164}$$

Here we have assumed the electric field to be zero; $v_\parallel$, and $v_\perp$ are the longitudinal and perpendicular to the direction of the magnetic field components of the particle velocity. The conservation of the transverse adiabatic invariant leads to the relationship between the transverse component of the particle velocity $v_\perp$ and the magnetic field $B(x)$: $v_\perp^2 = 2\mu mB$. Substituting this dependence in

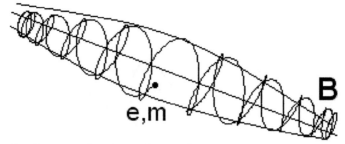

Fig. 3.7. Oscillations of a charged particle between two turning points in the magnetic field.

expression (3.164) we find

$$v_\parallel = \pm \left(\frac{2\,E}{m} - 2\mu m B(x) \right)^{1/2}. \tag{3.165}$$

In this expression the magnetic field $B(x)$ plays a role of effective potential. The particle turns back from the point where $B(x) = E/\mu m^2$, where the longitudinal component of the particle velocity vanishes: $v_\parallel = 0$.

Let us assume that the magnetic field has a minimum at some point and increases when $|x|$ grows as it is shown in Fig. 3.7.

In this case the particle oscillates between two turning points. The motion is periodic and the adiabatic invariant corresponding to this periodic motion is the longitudinal adiabatic invariant

$$J_\parallel = \frac{1}{2\pi} \oint p_\parallel \, dl. \tag{3.166}$$

Here integration is performed along the particle trajectory, that is along the magnetic field line since the particle moves along it.

● **Exercise.** Assuming that the magnetic field strength in equation (3.165) varies as $B(x) = B_0 \left(1 + (x/L)^2 \right)$, find the period of oscillations and the longitudinal adiabatic invariant of a charged particle, depending on its energy and transverse velocity at the point of the magnetic field minimum.

◇

If the distance between the turning points decreases the particle acquires the energy. We can estimate the value of the longitudinal adiabatic invariant as $J_\parallel \approx 2p_\parallel(t) L(t)/\pi =$ constant. As we see when $L(t)$ decreases, the longitudinal momentum $p_\parallel(t)$ increases. E. *Fermi* proposed this mechanism of acceleration of charged particles to explain acceleration of cosmic rays in the Galaxy.

3.4.2 Accuracy of Adiabatic Invariant Conservation

With the multiple time scale expansion we can find a solution to the equation

$$\ddot{x} + 2\nu\,(\varepsilon t)\,\dot{x} + \omega_0^2\,(\varepsilon t)\,x = 0 \tag{3.167}$$

for slowly varying damping rate and frequency ($\varepsilon, \nu \ll 1$). It is given by expression

$$x(t) = a(0)\left(\frac{\omega_0(0)}{\omega_0(t_1)}\right)^{1/2} \cos\left(\frac{1}{\varepsilon}\int^{\varepsilon t} \omega_0(\xi)\,d\xi\right) \exp\left(-\int^{\varepsilon t} \nu(\xi)\,d\xi\right). \qquad (3.168)$$

By direct inspection we see that the adiabatic invariant , $E(t)/\omega_0(t)$, is not conserved anymore. However, this is an example of behavior of a non-Hamiltonian system.

In the Hamiltonian systems with slowly varying parameters the adiabatic invariant is well conserved, but it is not so, of course, when the frequency of oscillations changes during the time comparable or shorter than the oscillation period. To illustrate this we discuss the case when the frequency $\omega_0(\varepsilon t)$ is discontinuous. We need to solve the equation

$$\ddot{x} + \omega_0^2(t)\,x = 0 \qquad (3.169)$$

with $\omega_0(\varepsilon t) = \chi(t)\Delta\omega + \omega_-$, where $\chi(t)$ is the Heaviside function

$$\chi(t) = \begin{cases} 1 & \text{when} & t > 0 \\ 0 & \text{when} & t < 0 \end{cases}. \qquad (3.170)$$

Matching of the solutions at $t = 0$, written in the form

$$x(t) = a_\pm \cos(\omega_\pm t) + b_\pm \sin(\omega_\pm t) \qquad (3.171)$$

in regions $t > 0$ and $t < 0$, gives the relationship between $a_\pm$ and $b_\pm$. Using this relationship we find that the adiabatic invariant value changes at $t = 0$ with

$$\Delta J = \frac{\Delta\omega}{2}\left(a_-^2 + b_-^2\frac{\omega_-}{\omega_+}\right) \neq 0. \qquad (3.172)$$

We see that the adiabatic invariant change is of the order of the change in the frequency.

When we discuss the accuracy of adiabatic invariant conservation we need to study the limit of *very slow* change of the frequency with time. The typical time of the frequency change is of the order of $T \propto \varepsilon^{-1}$ with $\varepsilon \ll 1$. In this limit the equation (3.169) can be formally considered as the Schrödinger equation. When $T \gg \omega^{-1}$ the equation can be solved with the WKB method.

From the courses of quantum mechanics (*L. D. Landau and E. M. Lifshits, Quantum Mechanics*) it is known that the amplitude of the over barrier reflection calculated in the frame of the WKB approximation is of the order of

$$\exp\left(-\frac{2}{\hbar}\int_{x_1}^{x_0} p\,dx\right). \qquad (3.173)$$

Here $x_1 \in (-\infty, +\infty)$ is real, x_0 is a point where $p(x_0) = 0$. In the generic case it can be a complex number. In our case the change of the adiabatic invariant is proportional to the amplitude of the reflected wave:

$$\Delta J \propto \exp\{-2\,Im \int_{t_1}^{t_0} \omega_0(t)dt\}, \qquad (3.174)$$

where t_0 is a complex number for which $\omega_0(t_0) = 0$. As a result the change of adiabatic invariant is of the order of $\exp(-1/\varepsilon_0)$, where $1/\varepsilon_0$ is proportional to the distance between points t_1 and t_0.

If we suppose that $\omega_0^2(t)$ is given by expression

$$\omega_0(t) = \frac{\omega_+ + \omega_-}{2} + \frac{\omega_+ - \omega_-}{2}\tanh\left(\frac{t}{T}\right), \qquad (3.175)$$

we find

$$t_0 = iT\,\text{Arctanh}\left(\frac{2\bar{\omega}}{\Delta\omega}\right) \equiv -i\alpha T. \qquad (3.176)$$

Here $\bar{\omega} = (\omega_+ + \omega_-)/2$ and $\Delta\omega = \omega_+ - \omega_-$. Hence the change of the adiabatic invariant is

$$\Delta J \approx \exp\left(-2\Im\left\{\int_{t_1}^{t_0=-i\alpha T}\left(2\bar{\omega} - \Delta\omega\tanh\left(\frac{t}{T}\right)\right)dt\right\}\right)$$

$$= \exp\left(-4\bar{\omega}T\left(\alpha - \frac{\Delta\omega}{2\bar{\omega}}\ln\cos\alpha\right)\right). \qquad (3.177)$$

The condition $\Delta J/J \ll 1$ requires

$$\max\{\bar{\omega}T, \Delta\omega T\} \gg 1. \qquad (3.178)$$

That is the change of the frequency must be slow ($\omega T \gg 1$) or $\dot{\omega}/\omega^2 \ll 1$.

3.4.3   Action-Angle Canonical Variables

In a generic case the adiabatic invariant is defined as

$$J = \oint p\frac{dq}{2\pi}. \qquad (3.179)$$

It is proportional to the value of area of the surface enclosed by a closed or periodic trajectory in a phase space. These two cases of closed and periodic trajectory are shown in Figs. 3.8a and 3.8b, respectively.

We can also say that the adiabatic invariant is, by definition, $(2\pi)^{-1}$ times the instantaneous area of the Hamiltonian $\mathcal{H}(q, t, \varepsilon t) = \mathcal{H}$ contour on which (q, p) lies. The adiabatic conservation of this value follows from two principles expressed in terms not of the individual particle motion, but of the continuum of particles moving around the contour $\mathcal{H}(q, t, \varepsilon t) = \mathcal{H}$ with different phases, which make a continuous train of particles in the phase plane.

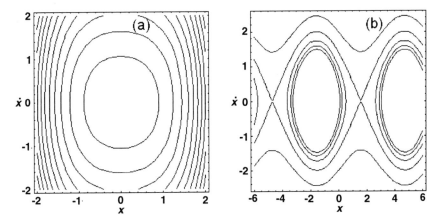

Fig. 3.8. Closed (a) and periodic (b) trajectory in the phase plane.

(i) The adiabatic principle. If the Hamiltonian change is slow enough the particles will all be affected in the same way and finally will be moving each after other around a new instantaneous $\mathcal{H}(q, t, \varepsilon t)$ =constant trajectory rather than being dispersed in some domain of the phase plane as it would be in a fast change.

(ii) Liouville's theorem. The area enclosed by any closed curve of particles in the phase space is conserved under any (either slow or fast) changing of the Hamiltonian.

The adiabatic invariance of the value of J, the surface enclosed by a trajectory of periodic motion, is the reason why the action-angle canonical variables are widely used in the Hamiltonian systems. In this case the canonical transformation is performed from coordinates p, q, where p is a momentum and q is a coordinate, to new coordinates J and θ, where J is the action, and θ is the angle. The canonical transformation is performed with the generating function $\mathcal{S}(J, q)$ as

$$p = -\frac{\partial \mathcal{S}(J, q)}{\partial q}, \qquad \text{and} \qquad \theta = \frac{\partial \mathcal{S}(J, q)}{\partial J}. \tag{3.180}$$

Calculating the change in function $\mathcal{S}(J, q)$ along one circuit, l_i, in a phase plane we find

$$\Delta_i \mathcal{S} = \oint_{l_i} p dq = 2\pi J_i. \tag{3.181}$$

The change in the angle θ is

$$\Delta_i \theta = \frac{\partial}{\partial J} \Delta_i \mathcal{S} = 2\pi \frac{\partial J_i}{\partial J}, \tag{3.182}$$

hence

$$\Delta_i \theta = 2\pi \delta_{ij}. \tag{3.183}$$

The new Hamiltonian, $\tilde{\mathcal{H}}$, in action-angle variables depends on J only. This gives the Hamilton equations in new variables:

$$\dot{J} = -\frac{\partial \tilde{\mathcal{H}}}{\partial \theta} = 0, \qquad \dot{\theta} = \frac{\partial \tilde{\mathcal{H}}}{\partial J} \equiv \omega(J). \tag{3.184}$$

Linear Oscillator

The Hamiltonian for motion of the system with one degree of freedom has the form

$$\mathcal{H}(p,q) = \frac{p^2}{2m} + U(q) = h. \tag{3.185}$$

For given value of $\mathcal{H}(p,q) = \mathcal{H}$ the momentum equal

$$p = (2m(h - U(q)))^{1/2}, \tag{3.186}$$

and the adiabatic invariant J is

$$J = \frac{1}{\pi} \int_{q_1}^{q_2} (2m(h - U(q)))^{1/2} \, dq. \tag{3.187}$$

For $U(q) = m\omega_0^2 q^2/2$, in the case of a linear oscillator, $q_1 = -q_2 = -(2\mathcal{H}/m\omega_0^2)^{1/2}$. Then the adiabatic invariant for the linear oscillator is equal to $J = \mathcal{H}/\omega_0$ or

$$\mathcal{H} = J\omega_0, \qquad \omega = \omega_0, \qquad \theta = \theta(0) + \omega_0 t. \tag{3.188}$$

Calculating the Hamiltonian in new canonical coordinates via the generating function $\mathcal{S}(J,q)$, we find

$$\mathcal{H}\left(\frac{\partial \mathcal{S}}{\partial q}, q\right) = \tilde{\mathcal{H}}(J) \tag{3.189}$$

with

$$\frac{\partial \mathcal{S}}{\partial q} = 2m\left(J\omega_0 - \frac{1}{2}m\omega_0^2 q^2\right)^{1/2}. \tag{3.190}$$

Finally we have relationships between the coordinate q, momentum p, and the action-angle coordinates, J and θ :

$$q = \left(\frac{2J}{m\omega_0}\right)^{1/2} \cos\theta, \qquad p = -\left(\frac{2J}{m\omega_0}\right)^{1/2} \sin\theta. \tag{3.191}$$

Action-Angle Variables for Particle Bouncing between Two Plates
Now we perform the canonical transformation of the coordinates from p, q to the action-angle coordinates, J, φ, for the Hamilton system with a phase plane shown in Fig. 3.9.

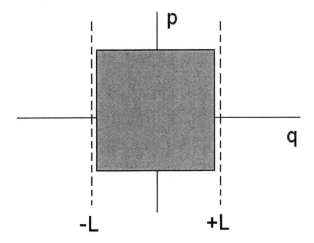

Fig. 3. 9. The phase plane of the particle bouncing between two plates.

In this case the action is proportional to the surface area enclosed by the trajectory of the particle with the energy E. It is equal to

$$J = 8\pi \left(2m \, EL\right)^{1/2}.$$ (3.192)

The Hamiltonian in the action-angle coordinates is

$$\tilde{\mathcal{H}}(J, \varphi) = \frac{(\pi J)^2}{8mL^2},$$ (3.193)

and the frequency of oscillations depends on J as

$$\omega\left(J\right) = \frac{\pi^2 J}{4mL^2},$$ (3.194)

since $\omega\left(J\right) = \partial\mathcal{H}/\partial J$. It increases with growths of J.

3.5 Nonlinear Oscillator

The problem of the pendulum. Action-angle coordinates for pendulum. Elliptic integrals. Elliptic functions of Jacobi. Motion of a charged particle in the vicinity of a null surface of the magnetic field. Phase anholonomy.

***

3.5.1 The Problem of the Pendulum

We write the equation of the pendulum motion in the form

$$\ddot{\theta} + \frac{g}{l}\sin\theta = 0,$$ (3.195)

where l is the length of the pendulum, g - the acceleration of gravity, θ - the angular displacement of the pendulum from its position of equilibrium. Normalizing time on $(l/g)^{1/2}$ and changing $\theta \to q$, we write

$$\ddot{q} + \sin q = 0. \tag{3.196}$$

We see that this is the Hamiltonian system with the Hamiltonian

$$H(p, q) = \frac{p^2}{2} - (1 + \cos q) = h, \tag{3.197}$$

where constant h is equal to zero for the separatrix. The constant h can be expressed as $h = -1 - \cos q_m$, with q_m being the maximum value of the angle at which $p = 0$.

From equation (3.196) we find

$$dt = \frac{dq}{2^{1/2}(\cos q - \cos q_m)^{1/2}} = \frac{d\phi}{(1 - \kappa^2 \sin^2 \phi)^{1/2}}. \tag{3.198}$$

We have introduced the transformation $\cos q = 1 - 2\kappa^2 \sin^2 \phi$, $\kappa = \sin(q_m/2)$. Therefore the time required for the pendulum to move from a position $q = 0$ at $t = 0$ to a position q is given by the integral

$$t = \int_0^\phi \frac{d\phi}{(1 - \kappa^2 \sin^2 \phi)^{1/2}}, \tag{3.199}$$

where the upper limit is obtained from the condition $\sin \phi = \kappa \sin(q/2)$. Expression (3.199) can be rewritten in terms of elliptic integrals: $t = F(\varphi, \kappa)$, where $F(\varphi, \kappa)$ is the elliptic integral of the first kind. The period of motion is equal to

$$T = 4 \int_0^{\pi/2} \frac{d\phi}{(1 - \kappa^2 \sin^2 \phi)^{1/2}} = 4K(\kappa). \tag{3.200}$$

Here $K(\kappa)$ is the complete elliptic integral of the first kind. For small amplitude of oscillations, $\kappa \ll 1$, expanding the integrand and integrating term by term, we find

$$T = 4 \int_0^{\pi/2} \left(1 + \frac{\kappa^2}{2} \sin^2 \phi + \frac{3\kappa^4}{8} \sin^4 \phi - \ldots \right) d\phi =$$

$$2\pi \left(1 + \frac{\kappa^2}{4} + \frac{9\kappa^4}{64} + \ldots \right) \tag{3.201}$$

Since $\kappa = \sin(q_m/2) = q_m/2 - q_m^3/48 + \ldots$, we have

$$T = 2\pi \left(1 + \frac{q_m^2}{4} + \frac{11 q_m^4}{3072} + \ldots \right) \tag{3.202}$$

We see that the nonlinear change of the period is proportional to the amplitude of oscillations.

To find the actual motion of the pendulum we need to inverse the integral (3.199). It yields

$$\text{sn}(t, \kappa) = \sin \phi = \frac{1}{\kappa} \sin \frac{q}{2}, \tag{3.203}$$

from which we finally obtain:

$$x = 2 \arcsin(\kappa \, \text{sn}(t, \kappa)) \quad \text{with} \quad \kappa = \sin(q_m/2) = (1 + h/2). \tag{3.204}$$

Action-Angle Coordinates for Pendulum

From equation (3.197) we see that the canonical momentum is $p = (2(h + 1 + \cos q))^{1/2} = (2(\cos q - \cos q_m))^{1/2}$. For action and angle variables we obtain

$$J(h) = \frac{2}{\pi} \int_0^{q_m} (2(h + 1 + \cos q))^{1/2} dq, \tag{3.205}$$

and

$$\theta = \partial J \partial h^{-1} \int_0^{q_m} \frac{dq}{(2(h + 1 + \cos q))^{1/2}}, \tag{3.206}$$

where $q_m = \arccos(-(1 + h))$ for libration, $h < 0$, i.e. $\kappa < 1$, and $q_m = \pi/2$ for rotation, $h > 0$, i.e. $\kappa > 1$. Expressions for the action and angle variables can be written in terms of elliptic integrals:

$$J = \frac{8}{\pi} \begin{cases} E(\kappa) - (1 - \kappa^2) K(\kappa), & \kappa < 1 \\ \frac{\kappa}{2} E\left(\frac{1}{\kappa}\right), & \kappa > 1 \end{cases}, \tag{3.207}$$

$$\theta = \frac{\pi}{2} \begin{cases} (K(\kappa))^{-1} F(\phi, \kappa), & \kappa < 1 \\ 2\left(K\left(\frac{1}{\kappa}\right)\right)^{-1} F\left(\frac{\phi}{2}, \frac{1}{\kappa}\right), & \kappa > 1 \end{cases}. \tag{3.208}$$

Using the relationship between the frequency, ω, and action, J, of periodic motion,

$$\frac{\partial J}{\partial h} = \frac{1}{\omega}, \tag{3.209}$$

we obtain

$$\omega = \frac{\pi}{2} \begin{cases} (K(\kappa))^{-1}, & \kappa < 1 \\ 2\kappa \left(K\left(\frac{1}{\kappa}\right)\right)^{-1}, & \kappa > 1 \end{cases}. \tag{3.210}$$

In the limit of small amplitude oscillations as $\kappa \to 0$ we find that $\omega = 1$; in dimensional units $\omega = (g/l)^{1/2}$. When $\kappa \to 0$ the trajectory approaches the separatrix in the phase plane (p, q). From (3.210) it follows that

$$\omega = \frac{\pi}{2} \begin{cases} 1/2 \ln\left(\frac{16}{(1 - \kappa^2)}\right), & \kappa < 1 \\ 1/\ln\left(\frac{16}{(1 - \kappa^2)}\right), & \kappa > 1 \end{cases}. \tag{3.211}$$

We see that in the vicinity of the separatrix the frequency tends to zero logarithmically.

Elliptic Integrals   Let us consider an ordinary differential equation

$$\left(\frac{dy}{dx}\right)^2 = a + by + cy^2. \tag{3.212}$$

We assume that the discriminant of the binomial equation in the right-hand side of (3.212), $\Delta = b^2 - 4ac$, is positive. Then the solution to equation (3.212) is

$$y = \begin{cases} \frac{\sqrt{\Delta}}{2|c|}\sin(\sqrt{|c|}x + \varphi), & \text{if} \quad c < 0 \\ \frac{\sqrt{-\Delta}}{2c}\sinh(\sqrt{c}x + \varphi), & \text{if} \quad c > 0 \end{cases}. \tag{3.213}$$

The natural generalization of equation (3.212) is the following

$$\left(\frac{dy}{dx}\right)^2 = a + by + cy^2 + dy^3 + ey^4. \tag{3.214}$$

If the roots of quartic equation are $\alpha, \beta, \gamma, \delta$, the right-hand side of this equation can be written as $e(y - \alpha)(y - \beta)(y - \gamma)(y - \delta)$. If, in particular $e = \kappa^2$, and the roots are $1, -1, 1/\kappa, -1/\kappa$, then we have

$$\left(\frac{dy}{dx}\right)^2 = (1 - y^2)(1 - \kappa^2 y^2). \tag{3.215}$$

For initial conditions $y(0) = 0$ and $y'(0) = 1$ the particular solution of (3.215) is the elliptic sine Jacobi:

$$y = \operatorname{sn}(x, \kappa) \tag{3.216}$$

Also one can have equations

$$\left(\frac{dy}{dx}\right)^2 = \frac{(1 - y^2)}{(1 - \kappa^2 y^2)} \tag{3.217}$$

and

$$\left(\frac{dy}{dx}\right)^2 = (1 + n^2 y^2)(1 - y^2)(1 - \kappa^2 y^2). \tag{3.218}$$

The elliptic integral of the first kind is defined as

$$F(x, \kappa) = \int_0^x \frac{dx}{((1 - x^2)(1 - \kappa^2 x^2))^{1/2}}, \qquad \kappa < 1 \tag{3.219}$$

or its equivalent

$$F(\phi, \kappa) = \int_0^\phi \frac{d\phi}{(1 - \kappa^2 \sin^2 \phi)^{1/2}}, \qquad \kappa < 1. \tag{3.220}$$

Here $x = \sin\phi$ and κ is called the modulus.

For $x = 1$ or $\phi = \pi/2$ we have $F(x = 1, \kappa) = F(\phi = \pi/2, \kappa) = K(\kappa)$. It is the complete elliptic integral of the first kind.

The elliptic integral of the second kind is defined as

$$E(x, \kappa) = \int_0^x \left(\frac{(1 - x^2)}{(1 - \kappa^2 x^2)} \right)^{1/2} dx, \qquad \kappa < 1 \qquad (3.221)$$

or its equivalent

$$E(\phi, \kappa) = \int_0^\phi (1 - \kappa^2 \sin^2\phi)^{1/2} d\phi, \qquad \kappa < 1. \qquad (3.222)$$

The complete elliptic integral of the second kind is $E(\kappa) = E(x = 1, \kappa) = E(\phi = \pi/2, \kappa)$.

Elliptic integral of the third kind is defined as

$$\Pi(x, n, \kappa) = \int_0^x \frac{dx}{((1 - nx^2)(1 - x^2)(1 - \kappa^2 x^2))^{1/2}}, \qquad \kappa < 1, \qquad (3.223)$$

and

$$\Pi(\phi, n, \kappa) = \int_0^\phi \frac{d\phi}{(1 + n\sin^2\phi)(1 - \kappa^2 \sin^2\phi)^{1/2}}, \qquad \kappa < 1. \qquad (3.224)$$

Elliptic Functions of Jacobi. The elliptic functions of Jacobi are defined as inverse of the elliptic integral of the first kind:

$$u = \int_0^\phi (1 - \kappa^2 \sin^2\phi)^{-1/2} d\phi, \qquad \mathrm{sn}(u, \kappa) = \sin\phi, \qquad \mathrm{cn}(u, \kappa) = \cos\phi,$$

$$\mathrm{dn}(u, \kappa) = (1 - \kappa^2 \sin^2\phi)^{1/2}, \quad \mathrm{am}(u, \kappa) = \phi, \quad \mathrm{tn}(u, \kappa) = \frac{\mathrm{sn}(u, \kappa)}{\mathrm{cn}(u, \kappa)} = \tan\phi.$$

$$(3.225)$$

3.5.2 Motion of Charged Particle in the vicinity of a Null Surface of the Magnetic Field

We consider the configuration with a uniform electric field and magnetic field reversing its sign on the $x-$ axis. The fields are given by

$$\mathbf{E} = -E_0 \mathbf{e}_x, \qquad \mathbf{B} = -by \mathbf{e}_z. \qquad (3.226)$$

Far from the axis, the Larmor radius of the particle is much less than its distance from the axis. In this region the Larmor frequency ω_B must be calculated for the value of the magnetic field in the region where the particle moves. In the vicinity of the axis where the magnetic field vanishes, the charged particle motion is localized in the region with the size larger than the local value of the Larmor radius.

The electron trajectory is given by a solution of equations:

$$\ddot{x} = \frac{eE}{m} + \frac{eb}{mc}\dot{y}y, \tag{3.227}$$

$$\ddot{y} = -\frac{eb}{mc}\dot{x}y. \tag{3.228}$$

Here and below a dot denotes the time derivative.

Integrating equation (3.227) with respect to time we obtain a relationship

$$\dot{x} = v_0 + \frac{eE}{m}t + \frac{eb}{2mc}(y^2 - y_0^2). \tag{3.229}$$

Here v_0 and y_0 are the initial values of the x-component of the particle velocity and the y-coordinate, respectively. Inserting this expression for $\dot{x}$ into equation (3.228), we find the latter to have a Hamiltonian form with a time-dependent Hamiltonian. It is equal to

$$\begin{aligned} H(p_y, y, t) &= \frac{p_y^2}{2m} + V(y, t) \\ &= \frac{p_y^2}{2m} + \frac{eby^2}{2m^2c}P(t) + \frac{e^2b^2y^4}{8m^2c^2} \\ &= \frac{p_y^2}{2m} + \frac{eby^2}{2mc}\left(v_0 + \frac{eE}{m}t - \frac{eby_0^2}{2mc}\right) + \frac{e^2b^2y^4}{8m^2c^2}. \end{aligned} \tag{3.230}$$

Depending on the sign of the function

$$P(t) = m\left(v_0 + \frac{eE}{m}t - \frac{eb}{2mc}y_0^2\right) \tag{3.231}$$

the effective potential, $V(y, t)$, has one or two local minima. It has one minimum for positive P and two minima when P is negative.

In the case when the electric field vanishes, the function P given by equation (3.231) is independent of time. When P is negative, the effective potential $W(y)$ has two minima at $y = y_+$ and $y = y_-$. For negative values of the Hamiltonian, $H(p_y, y) = h$, the particle is trapped in the region far from the surface $y = 0$. The trajectory is localized either near the surface $y = y_+$ or near $y = y_-$, where subscripts $\pm$ stand for the upper half plane $y > 0$ and the negative half plane $y < 0$, respectively. We shall say the particle to be in the outer region. Physically, this regime corresponds to the case when the Larmor radius of the particle is much less than the distance from the neutral surface. For sufficiently small h, the particle motion in the first approximation is the gyration around the magnetic field line with a frequency which is equal to the local value of the Larmor frequency, $\omega_B = eby_\pm/mc$. Due to the magnetic field gradient, the center of the Larmor circle moves on the surface $y = y_\pm$ in the negative direction along the x-axis; we have for the coordinates $y_\pm$: $y_\pm(P) \approx \pm(2|P|c/eb)^{1/2}$.

When $P < 0$ and $h > 0$, the Larmor radius is larger than the distance from the neutral surface and the particle trajectory meanders in the vicinity of the neutral surface.

For a positive P, the particle trajectory is in the inner region near the neutral surface. If the Hamiltonian value is small, $h < 2\,P/m$, the particle moves along the x-axis with the velocity $v_0 - eby_0^2/2mc$ and oscillates in the transverse direction. The frequency of oscillations is equal to $\Omega = (eb\,P/m^2c)$.

In a uniform electric field directed along the x-axis, perpendicularly to the magnetic field, the function P is a function of time according to equation (3.231). As a result the position of the local minimum, $y_\pm$, and the gyration frequency, $\omega_B(y_\pm)$, obtained above, change with time. In the outer region far from the neutral surface, $y_\pm$ and ω_B change gradually with time. The latter corresponds to the $\mathbf{E} \times \mathbf{B}$ drift of the particle in the y-direction which can be outward or inward depending on the relative signs of the electric and magnetic field. Below, the electric field is assumed to be negative. The particle trajectory in the inner region can be expressed in terms of the Airy functions.

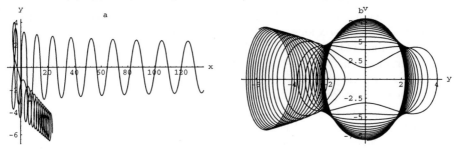

Fig. 3. 10. Motion of charged particle in the vicinity of null line of the magnetic field in a homogeneous electric field: (a) trajectory in (x, y) plane; (b) trajectory in the phase plane $(\dot{y} \equiv v, y)$.

In Fig. 3.10 the particle trajectory is shown for $E \neq 0$. We normalize the space and time variables on $Y = (mc^2\,E/eb^2)^{1/3}$, and $T = (m^2c/e^2\,Eb)^{1/3}$. As a result, the velocity and the function P are normalized on $Y/T = (e\,E^2c/mb)^{1/3}$ and mY/T, respectively.

Using the asymptotic representation of the Airy function for $x \to +\infty$ we obtain, in dimensional units, the time dependence of the y-coordinate

$$y \approx y_0 \left(\frac{1}{\Omega_V(t)T}\right)^{1/4} \sin\left(\int_0^t \Omega_V(t)\,dt + \frac{\pi}{4}\right), \qquad (3.232)$$

with $\Omega_V(t)$ being the instantaneous frequency of oscillations.

$$\Omega_V(t) = \left(\frac{e^2\,Ebt}{m^2c}\right)^{1/2}, \qquad (3.233)$$

and corresponds to that given by expression for Ω where the function P has been replaced by eEt.

Now, we calculate the energy of transverse motion

$$E_{\perp \text{in}}(t) = m< \dot{y}^2 + \Omega_V^2 y^2 >/2 = (my_0^2/2)\Omega_V(t)T. \qquad (3.234)$$

We see that the adiabatic invariant $J_{\text{in}} = E_{\perp \text{in}}(t)/\Omega_V(t) = (my_0^2/2)T$, associated with the transverse motion in the inner region, is conserved.

It is well-known that in the outer region far from the neutral surface, there is a conservation of the adiabatic invariant which equals the magnetic flux through the surface enclosed by the Larmor circle. This yields an expression for the adiabatic invariant associated with the particle motion along the y-coordinate, $J_{\text{out}} = E_{\perp \text{out}}(t)/\omega_B(t)$, where $E_{\perp \text{out}}(t) =< \dot{y}^2 + \omega_B^2 y^2 > /2$ and $\omega_B(t)$ is the instantaneous value of the Larmor frequency. To match the solutions in the outer and inner regions we assume the parameter $\varepsilon_c = (e\, E^2 c/mbv^3)^{1/2}$ to be small. Here the particle velocity v is taken at the instant when the particle trajectory crosses the boundary between the inner and outer regions.

The boundary between the outer and inner regions corresponds to the separatrix in the phase plane $(y, \dot{y})$. A generic form of the separatrix for the Hamiltonian given by equation (3.230), with the time dependence assumed to be frozen, is of a figure eight. It divides the phase plane into three regions. Two of them are inside the separatrix lobes, corresponding to the particle motion near one of the local minima of the effective potential, i.e. to the motion in the outer region. The evolution of the particle trajectory in the phase plane is described by the Hamilton equations with the Hamiltonian $H(p_y, y, \varepsilon_c t)$ (3.230), where the time dependence is slow for small ε_c. The adiabatic invariant value is equal to the area of the surface enclosed by the trajectory in the phase plane. Far from the separatrix, the accuracy of the adiabatic invariant conservation is exponential. When the trajectory crosses the separatrix, the adiabatic invariant changes.

To determine the adiabatic invariant change for small ε_c we can apply the theoretical results discussed above, where it has been demonstrated that during the separatrix crossing, the adiabatic invariant acquires a value equal to the area of the surface enclosed by the separatrix divided by 2π. The accuracy of the adiabatic invariant conservation is of the order of $\varepsilon_c \ln \varepsilon_c$. The particle trajectory in the phase plane $(y, \dot{y})$ is shown in Fig.3.10(b).

We observe the change of the particle motion during the separatrix crossing. The adiabatic invariant increases twice because during such a separatrix crossing, the area enclosed by the trajectory increases by two times, and we obtain the relationship between the values of the adiabatic invariants in the outer and inner regions as $J_{\text{in}} = 2J_{\text{out}} + o(\varepsilon_c \ln \varepsilon_c)$.

3.6   Phase Anholonomy

Above we discussed the behavior of the adiabatic invariants which are related to the action variables. Now we can ask what is the behavior of the angle-coordinate conjugated to the action under an adiabatically slow change of the

Hamiltonian assuming that the dynamic system is carried slowly around the closed cycle. As we have seen, the instantaneous frequency for frozen dependence of the Hamiltonian on time is given by $\partial H/\partial J$. That is why we would write for the phase change during time t

$$\Delta\theta = \int_0^t \frac{\partial H(J,t)}{\partial J} dt. \tag{3.235}$$

However, this formula is incomplete because the angle changes also by virtue of the changing of the (J,θ) coordinate system in phase space. To reveal the full change of the phase along the path in the parameter space we must consider the time dependence of the Hamiltonian $H(p,q,\mathbf{X}(t))$, where the vector $\mathbf{X}(t)$ gives a path in the parameter space, assuming $\mathbf{X}(t)$ executes a loop.

In the angle-action coordinates the change of an action and angle are given by the Hamilton equations:

$$\dot{J} = -\frac{\partial H}{\partial \theta} + \mathbf{X}(t)\frac{\partial J}{\partial \mathbf{X}(t)}, \tag{3.236}$$

$$\dot{\theta} = \frac{\partial H}{\partial J} + \mathbf{X}(t)\frac{\partial \theta}{\partial \mathbf{X}(t)}. \tag{3.237}$$

The first term in the right hand side of equation (3.236) vanishes by virtue of the choice of the action-angle coordinates; the last terms in the right hand side 3.236 are the rates of change of action and angle at a fixed point (J,θ) in the phase space. For nonadiabatic dependence of the Hamiltonian on time these equations lead to changes in both J and θ, which depend on initial values of both J and θ. To find the adiabatic excursion we average equations (3.236) and (3.237) around the closed contour $H(p,q,\mathbf{X}(t))$ =const, for frozen dependence of the Hamiltonian on time. Since the contour $H(p,q,\mathbf{X}(t))$ =const is specified by the value of the adiabatic invariant we define the averaged function of action J by

$$\langle f(J) \rangle = \frac{1}{2\pi} \oint f d\theta \equiv \frac{1}{2\pi} \int f(q,p)\delta(J(q,p) - J)dqdp, \tag{3.238}$$

where $\delta(x)$ is the Dirac delta-function. The average in equation (3.236) vanishes identically according to the Liouville theorem and gives $\dot{J} = 0$. Instead, after the averaging, the right-hand side of equation (3.237) is not equal to zero. Integrating equation (3.237) with respect to time we find that the change in the conjugate to action coordinate θ is given by

$$\Delta\theta = \oint \frac{\partial H}{\partial J} dt + \oint \left\langle \frac{\partial \theta}{\partial \mathbf{X}} \right\rangle d\mathbf{X}. \tag{3.239}$$

The first term is the integral of the local rate of change and may be called the "dynamic" angle. The second is the geometric angle, Hannay's angle

(*J.H.Hannay, 1985*). It is an anhalonomy or non-integrability of the angle variable under adiabatic excursions of the Hamiltonian. The value of Hannay's angle depends on the path $\mathbf{X}(t)$ in the external parameter space. Since the coordinate θ may be chosen somewhat arbitrarily at each $\mathbf{X}(t)$, the change in an angle, $\Delta\theta$, is defined only for cyclic change when $\mathbf{X}(\mathbf{t})$ returns to its initial value. Therefore Hannay's angle is in general nonvanishing when the path $\mathbf{X}(t)$ through parameter space is closed, taking the functional form of the Hamiltonian (but not the actual dynamical state) back to the starting point.

Example. To illustrate the geometrical angle holonomy we consider the rotation of a curve in real space. Let the curve to be a "hoop" of area A and perimeter L in a plane with a bead of unite mass moving without any friction around it. The hoop is slowly turned through one revolution. If Ω is the hoop's angular velocity and $\mathbf{q}(t)$ the position of the bead in the hoop frame, the canonical momentum $\mathbf{p}(t)$ is given by

$$\mathbf{p} = \dot{\mathbf{q}} + \mathbf{\Omega} \times \mathbf{q}. \tag{3.240}$$

The action is the integral of $\mathbf{p}$ around the hoop

$$J = \frac{1}{2\pi} \oint \mathbf{p} \cdot d\mathbf{q} = \frac{1}{2\pi} \left(\oint \dot{\mathbf{q}} d\mathbf{q} + 2\Omega A \right). \tag{3.241}$$

One can say that the adiabatically conserved quantity is the spatial average speed of the bead around the hoop plus $\Delta v = 2\Omega A/L$. If Ω rises from zero and falls back slowly enough with time interval 2π, that is during one revolution of the hoop, then the time integral of the averaged speed is $\Delta s = -4\pi A/L$ greater than it would have been had Ω remained zero. Thus the extra angle change is

$$\Delta\theta = \frac{2\pi\Delta s}{L} = -\frac{8\pi^2 A}{L^2}. \tag{3.242}$$

This reduces the change of the angle -2π, when the hoop rotates catching up on the bead by one revolution, so that the bead has made one fewer turns in the hoop frame.

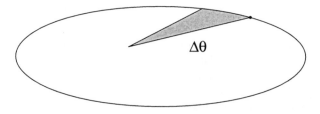

Fig. 3.11. Change in the angle during the bead motion along the hoop.

•

Chapter 4

Behavior of a Dynamic System near Separatrix

Change of adiabatic invariants in the crossing of the separatrix. Motion in the vicinity of the separatrix. Change of the adiabatic invariant of nonlinear pendulum during the separatrix crossing. Behavior of a dynamic system near a separatrix in the phase plane (The Melnikov's theory). Change of adiabatic invariant during the separatrix birth. Nonlinear resonances, their overlap, standard mapping. Stochastic layers. Strange attractor. The Lorenz system.

***

4.1 Change of Adiabatic Invariants in the Crossing of the Separatrix

The separatrix crossing by the particle trajectory is illustrated in Fig. 4.1.

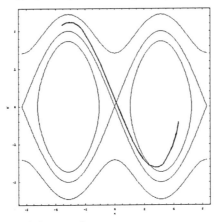

Fig. 4.1. The separatrix crossing.

S. BULANOV

The surface enclosed by the trajectory before the crossing is equal to J_0 , and that after the crossing is a sum of two values, J_1 and J_2. Had the adiabatic invariant been conserved, we would have had

$$J_0 = J_1 + J_2. \tag{4.1}$$

However due to the break of adiabatic approximation near the separatrix there is a relationship

$$J_1 + J_2 \doteq J_0 + \Delta J, \tag{4.2}$$

with ΔJ being a finite change in the adiabatic invariant value. Our goal is to find the relative change of the adiabatic invariant: $\Delta J / J$ (*Timofeev, 1978, Neishdadt, 1984*).

On the separatrix the period of particle motion is equal to infinity, so the parameter of adiabaticity becomes large:

$$\frac{\dot{\omega}}{\omega^2} \to \infty. \tag{4.3}$$

The critical point on the separatrix is the X-point (saddle point; see Fig. 4.2).

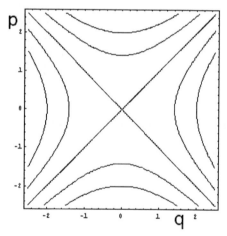

Fig. 4.2. X-point in the phase plane.

In the neighborhood of the saddle point we can use a local approximation for the Hamiltonian. It leads to the representation of $H(p, q)$ as it follows:

$$H(p, q) \approx \frac{p^2 - q^2}{2} = h. \tag{4.4}$$

Here the value h marks the trajectory, and $h = 0$ on the separatrix.
The Hamilton equations

$$\dot{p} = -\frac{\partial H}{\partial q}, \qquad \dot{q} = \frac{\partial H}{\partial p} \tag{4.5}$$

give

$$\dot{p} = q, \qquad \dot{q} = p \tag{4.6}$$

with a solution

$$p(t) = p(0)\cosh t + q(0)\sinh t, \qquad q(t) = q(0)\cosh t + p(0)\sinh t. \tag{4.7}$$

According to expression (4.4) the Hamiltonian equals $\left(p^2(0) - q^2(0)\right)/2 = h$. On the separatrix, where $h = 0$, we find $|p(0)| = |q(0)|$, that gives

$$\begin{aligned} p = p(0)\exp(t), \quad & q = q(0)\exp(t\ ) \quad \text{for} \quad p(0) = q(0), \\ p = p(0)\exp(-t), \quad & q = q(0)\exp(-t) \quad \text{for} \quad p(0) = -q(0). \end{aligned} \tag{4.8}$$

Near the separatrix, where $h \ll 1$, we can estimate the time that a particle spends in the neighborhood of the X-point as

$$t \approx \ln(1/h). \tag{4.9}$$

Now we find the value of the change of the particle energy while it moves near the X-point assuming that time dependent perturbation of the Hamiltonian has a form

$$H(p, q, t) = H_0(p, q) + \varepsilon V(q, t). \tag{4.10}$$

Since

$$\dot{H} = \{H_0, H\} = -\varepsilon p \frac{\partial V}{\partial q}, \tag{4.11}$$

where the Poisson brackets are

$$\{f, g\} = \frac{\partial f}{\partial p}\frac{\partial g}{\partial q} - \frac{\partial f}{\partial q}\frac{\partial g}{\partial p}, \tag{4.12}$$

the change of the particle energy is

$$\Delta E = -\varepsilon \int p \frac{\partial V}{\partial q} dt. \tag{4.13}$$

That is $\Delta E \approx \varepsilon$, which gives $h \approx \varepsilon$. As a result expression (4.9) can be rewritten as

$$t \approx \ln(1/\varepsilon). \tag{4.14}$$

This in turn gives for the change of the adiabatic invariant

$$\Delta J \approx p \cdot \Delta q \approx q \cdot \Delta p \approx t\Delta E \approx \varepsilon \ln \varepsilon. \tag{4.15}$$

If in the vicinity of critical points the Hamiltonian can not be approximated in the quadratic form similar to that given by equation (4.4), then we cannot use estimation (4.15) for the invariant change. For the Hamiltonian

$$H = \frac{p^2}{2} + V(q, \varepsilon t), \tag{4.16}$$

with $V(q) \propto q^n$, $n > 2$ near $q = 0$, the period of motion is not logarithmic:

$$T = \oint \frac{dq}{(2(h - V(q)))^{1/2}} \cong h^{(2-n)/2n}. \tag{4.17}$$

The proper estimate for the adiabatic invariant change gives

$$\Delta J \simeq \varepsilon^{(2+n)/2n}. \tag{4.18}$$

This dependence corresponds to the case when several saddle points merge and to the rise or disappearance of the region of the finite motion in the phase space and, in particular to the birth of the separatrix. The case of separatrix birth is determined by the semicubic cusp point, where $p \propto q^{3/2}$. The last corresponds to $n = 3$ and the adiabatic invariant change is of the order of $\Delta J \simeq \varepsilon^{5/6}$.

More careful calculations may be done taking into account the behavior of a trajectory near the separatrix (see Fig.4.3).

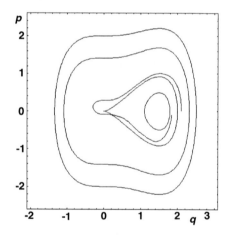

Fig. 4.3. Trajectory near the separatrix.

The adiabatic invariant value must be calculated as

$$J = \frac{1}{2\pi} \int_1^2 p\,dq = \frac{1}{2\pi} \left(\oint_S p_0\,dq_0 + \int_1^2 (p_1\,dq_0 + p_0\,dq_1) \right). \tag{4.19}$$

In the first term in the right-hand side integration is performed along the unperturbed separatrix S. That gives a value of J_0. The second term is equal to ΔJ which is nothing more than the change of the adiabatic invariant.

4.2 Motion in the Vicinity of the Separatrix

In this section we estimate the change of the adiabatic invariant for the case of one X-point. The system under consideration has the Hamiltonian of the form

$$H = \frac{p^2}{2} - (1 + \varepsilon t)\left(\frac{1}{2}q^2 - \frac{1}{4}q^4\right) = h. \qquad (4.20)$$

The potential in this case has two minima. The location of the separatrix corresponding to $h = 0$ depends on the time εt, that leads to changes of the potential shape and depth.

Let us introduce a new function $\gamma^2(\varepsilon t) = 1 + \varepsilon t$, and a new time variable

$$\eta = \frac{1}{\varepsilon}\int^{\varepsilon t}\gamma(s)ds. \qquad (4.21)$$

Under these transformations the equation of motion takes the form

$$q'' - q + q^3 = -\frac{\gamma'}{\gamma}q', \qquad (4.22)$$

where the prime denotes a derivative with respect to η. The advantage of this transformation is that the obtained equation in the zeroth order on the parameter ε value describes the motion in frozen potential with a stationary separatrix. The non-stationarity of the Hamiltonian is described by the effective friction term in the right-hand side of equation (4.22) which is of order $\gamma'/\gamma \approx \varepsilon/2 \ll 1$ for $\varepsilon \ll 1$. Another remarkable circumstance is that the change of the coordinates inverse to that which is given by equation (4.21) transforms the dissipative system to a Hamiltonian one.

The solution of zeroth order in ε which corresponds to the motion on the separatrix for the Hamiltonian for $\varepsilon t = 0$ obeys the equation

$$\left(q^{(0)}\right)' = \left(\left(q^{(0)}\right)^2/2 + \left(q^{(0)}\right)^4/4\right)^{1/2}, \qquad (4.23)$$

and is given by the expressions

$$q^{(0)} = \frac{2^{1/2}}{\cosh\eta}, \qquad \left(q^{(0)}\right)' = -\frac{2^{1/2}\sinh\eta}{\cosh^2\eta}. \qquad (4.24)$$

Now we seek the first order solution in ε in the form of a series

$$q = q^{(0)} + \varepsilon q^{(1)} + \cdots. \qquad (4.25)$$

The problem under consideration is reduced to the solution of equations with the Hamiltonian given by equation (4.20). They correspond to the Hamiltonian H_0 in the zeroth order in ε. The solution $(q^{(0)}, p^{(0)})$, governed by equations

(4.24), for $p^{(0)} \equiv (q^{(0)})'$ corresponds to the energy $h = 0$ in this case. The equation of motion in zeroth order on ε may be written as

$$\left(q^{(0)} \right)'' = F \left(q^{(0)} \right) . \tag{4.26}$$

In the first order in ε we have

$$\left(q^{(1)} \right)'' = \frac{\partial F}{\partial q^{(0)}} q^{(1)} + \varepsilon f(q^{(0)}, \left(q^{(0)} \right)', \eta) , \tag{4.27}$$

where εf is a perturbation.

Considering the term in the right-hand side of equation (4.27) as a perturbation we need to find the perturbed solution to the second order ordinary differential equation.

Corollary. *Perturbed Solution to the Second Order Ordinary Differential Equation.* We shall solve the problem for equation

$$\frac{d^2 x}{dt^2} = f^0(x) + \varepsilon f^1 \left(x, \frac{dx}{dt}, t \right) \tag{4.28}$$

with $\varepsilon \ll 1$.

We suppose the zeroth order solution in the parameter ε to be known. It is $x^{(0)}(t)$, and it obeys equation

$$\frac{d^2 x^{(0)}}{dt^2} = f^0 \left(x^{(0)} \right) . \tag{4.29}$$

To find the solution to first order in the parameter ε, that is $x^{(1)}(t)$, we need to solve equation

$$\frac{d^2 x^{(1)}}{dt^2} = \frac{\partial f^0}{\partial x} \Bigg|_{x = x^{(0)}} x^{(1)} + \varepsilon f^1 \left(x^{(0)}, \frac{dx^{(0)}}{dt}, t \right) . \tag{4.30}$$

Differentiation of equation (4.29) with respect to time gives

$$\frac{d^3 x^{(0)}}{dt^3} = \frac{\partial f^0}{\partial x} \Bigg|_{x = x^{(0)}} \frac{dx^{(0)}}{dt} . \tag{4.31}$$

From equations (4.30) and (4.31) it follows

$$\frac{d^2 x^{(1)}}{dt^2} \frac{dx^{(0)}}{dt} = \frac{d^3 x^{(0)}}{dt^3} x^{(1)} + \varepsilon f^1 \left(x^{(0)}, \frac{dx^{(0)}}{dt}, t \right) \frac{dx^{(0)}}{dt} . \tag{4.32}$$

Then we add a term $(d^3 x^{(0)}/dt^3)(dx^{(1)}/dt)$ into the right-hand side and the left-hand side. As a result we obtain equation

$$\frac{d^2 x^{(1)}}{dt^2} \frac{dx^{(0)}}{dt} + \frac{d^3 x^{(0)}}{dt^3} \frac{dx^{(1)}}{dt}$$

$$= \frac{d^3x^{(0)}}{dt^3}x^{(1)} + \frac{d^3x^{(0)}}{dt^3}\frac{dx^{(1)}}{dt} + \varepsilon f^1\left(x^{(0)}, \frac{dx^{(0)}}{dt}, t\right)\frac{dx^{(0)}}{dt} \qquad (4.33)$$

with a solution

$$\frac{dx^{(0)}}{dt}\frac{dx^{(1)}}{dt} = \frac{d^2x^{(0)}}{dt^2}x^{(1)} + \varepsilon \int_0^t f^1 \frac{dx^{(0)}}{dt}dt + C_1. \qquad (4.34)$$

Here C_1 is an arbitrary constant. From equation (4.34) we find

$$x^{(1)} = \dot{x}^{(0)}\left(\int_0^t \frac{dt'}{\left(\dot{x}^{(0)}\right)^2}\left(\varepsilon \int_0^{t'} f^1 \, \dot{x}^{(0)} dt'' + C_1\right) + C_2\right), \qquad (4.35)$$

with C_2 being a second arbitrary constant.

★

Using expression (4.35) and the zero order solution (4.24) we find

$$q^{(1)}(\eta) = \left(q^{(0)}\right)'(\eta)\left(C + \int_0^\eta \frac{d\tilde{\eta}}{\left(\left(q^{(0)}\right)'\right)^2}\left(\varepsilon \int_0^{\tilde{\eta}} f^1 \left(q^{(0)}\right)' d\tilde{\eta} + h\right)\right), \qquad (4.36)$$

where the constant C is proportional to $q^{(1)}(0)$.

If the integral $\int_0^\eta f^1 \left(q^{(0)}\right)' d\tilde{\eta}$ converges for $\eta \to \pm\infty$ and its value is equal to $M_\pm$ respectively, then for $h = -\varepsilon M$ the expression (4.36) is needed to be evaluated carefully. Namely, if the function f is such that

$$\varepsilon \int f^1 q'_0 d\eta \approx \varepsilon M - C_3 \exp(-\eta), \qquad (4.37)$$

where C_3 is constant, then we have

$$q^{(1)}(\eta) \approx C\,C_4\,exp(-\eta) + \frac{C_3}{C_4}. \qquad (4.38)$$

Here we use the fact that the solution $q^{(0)}$ can be approximated as

$$q^{(0)}(\eta) \approx C_4 \exp(\pm\eta) \qquad (4.39)$$

for $\eta \to \pm\infty$. In particular, in the case corresponding to (4.24) the constant is $C_4 = 2\sqrt{2}$. From equation (4.38) it follows that for $\eta \to \pm\infty$ $q^{(1)}$ tends to the constant C_3/C_4. These solutions describe splitting of the separatrix under the perturbation into two branches with the values of energy $h_\pm = -\varepsilon M_\pm$ and shift of a singular point to the value C_3/C_4^2 . Below the separatrix means the unperturbed separatrix, whose position is calculated for a frozen dependence of the Hamiltonian on time.

Substituting the expressions given by equation (4.24) into equation (4.36) we find the solution of the first order in ε for the Hamiltonian (4.20). For the initial conditions

$$q^{(1)}(0) = \frac{h}{2^{1/2}}, \qquad \left(q^{(1)}\right)'(0) = 0, \tag{4.40}$$

it has the form

$$q^{(1)}(\eta) = \varepsilon \frac{2^{1/2}}{12} \frac{\sinh^3 \eta}{\cosh^2 \eta} + h \frac{2^{1/2}}{4} \left(\frac{3}{\cosh \eta} - \cosh \eta - 3\eta \frac{\sinh \eta}{\cosh^2 \eta} \right). \tag{4.41}$$

The particle crosses the separatrix for the variable η value equal to η^*, such that $q^{(1)}(\eta^*) = 0$ and $\left(q^{(1)}\right)'(\eta^*) = 0$. The crossing takes place if $|h| < \varepsilon/3$. This gives an estimate for the width of the non-adiabatic region.

We estimate the adiabatic invariant value as a function of time which is presented by an area enclosed by the trajectory, and using the obtained solutions. The area enclosed by a perturbed trajectory between the points η_+ and η_-, for which $q' = \left(q^{(0)}\right)' + \left(q^{(1)}\right)' = 0$, divided by π, can be represented as

$$J = \frac{1}{\pi} \int p\,dq = J^{(0)}(\varepsilon t) + J^{(1)}, \tag{4.42}$$

where $J^{(0)}$ stands for the area enclosed by the separatrix and $J^{(1)}$ is due to the perturbation $q^{(1)}$.

As a function of η, $J^{(0)}$ is constant, i.e. the separatrix is immobile in terms of η. But $J^{(0)}$ is dependent on t:

$$J^{(0)} \cong \frac{1}{\pi} \left(\int_{-\infty}^{+\infty} \left(\left(q^{(0)}\right)' \right)^2 dt + \frac{\varepsilon}{2} \int_{-\infty}^{+\infty} t \left(\left(q^{(0)}\right)' \right)^2 dt \right). \tag{4.43}$$

For $J^{(1)}$ we have:

$$J^{(1)} = \frac{1}{\pi} \int \left(p^{(0)} dq^{(1)} + p^{(1)} dq \right) = \frac{2}{\pi} \int_{\eta-}^{\eta+} \left(q^{(1)}\right)' \left(q^{(0)}\right)' d\eta$$

$$= \left(\frac{\varepsilon}{6\pi} \left(\frac{1}{\cosh^4 \eta} - \frac{7}{\cosh^2 \eta} - 10\ln(\cosh \eta) \right) \right.$$

$$\left. + \frac{h}{\pi} \left(\eta - \tanh \eta + \frac{3}{2} \tanh^3 \eta \right) \right) \Big|_{\eta_-}^{\eta_+} \tag{4.44}$$

where

$$\eta_\pm = \pm \frac{1}{2} \ln \left(\frac{48}{\varepsilon \pm 3h} \right). \tag{4.45}$$

The solution $q^{(0)}(\eta) + q^{(1)}(\eta)$ has the following asymptotes as η tends to $\pm\infty$

$$q_{\mp}(\eta) = \begin{cases} 2^{3/2} \exp(\eta) - \dfrac{1}{2^{5/2}} \left(h + \dfrac{\varepsilon}{3} \right) \exp(-\eta), & \eta \to -\infty, \\[2mm] 2^{3/2} \exp(-\eta) + \dfrac{1}{2^{5/2}} \left(h - \dfrac{\varepsilon}{3} \right) \exp(\eta), & \eta \to +\infty. \end{cases} \tag{4.46}$$

Matching of the solution asymptotes for each step of motion near the separatrix gives the changes of the energy and a dimensionless time interval $\Delta\eta$ during which this step is accomplished

$$h_n = h_0 - \frac{2}{3}\varepsilon n, \qquad \Delta\eta = \ln\left(\frac{48}{3h_n - \varepsilon}\right). \qquad (4.47)$$

For the energy magnitudes $h = \pm\varepsilon/3$ the duration of one step becomes equal to infinity, as it follows from equation (4.47). From the dependences given by equation (4.46) we see that the trajectories asymptotically approach the coordinate origin on a phase space, that is on the coming into the saddle point vicinity and outgoing from the saddle point vicinity to the separatrix branches. These branches appear as a result of a perturbation and are described by equation (4.38) for $h_\pm = -\varepsilon M_\pm$. In this case $\varepsilon M_\pm = \mp\varepsilon/3$.

The summation over n- steps leads to the following value of the invariant in the n-th step

$$\begin{aligned}
J_n \; = \; & J_0 + \frac{1}{\pi}\left(h_n \ln\frac{48}{(9h_n^2 - \varepsilon^3)^{1/2}} + \frac{5}{6}\varepsilon \ln\left(\frac{3h_n + \varepsilon}{3|h_n| - \varepsilon}\right)\right. \\
& \left. + h_n + \frac{1}{3}\varepsilon\sum_{k=-n}^{k=+n} \ln\left(\frac{48}{3h_k - \varepsilon}\right)\right). \qquad (4.48)
\end{aligned}$$

Here $J_0 = 8/\pi$ is the area enclosed by the separatrix at $t = 0$.

4.3 Change of the Adiabatic Invariant of Nonlinear Pendulum During the Separatrix Crossing

Now we calculate the change of the adiabatic invariant of nonlinear pendulum (*Timofeev, 1978*). We write the modified equation for the pendulum motion as

$$\ddot{q} + A(t)\sin q = 0, \qquad (4.49)$$

where $A(t)$ is assumed to be a slow function of time of the form $A(t) = 1 + \varepsilon t$ with $\varepsilon \ll 1$. For unperturbed motion, $\varepsilon t = 0$, it follows that the separatrix trajectory, $h = 0$, is given by

$$q^{(0)}(t) = 4\text{Arctanh}\exp(t) - \pi. \qquad (4.50)$$

Using expression (4.35), we calculate perturbation of the trajectory for both $\varepsilon t \neq 0$, and $h \neq 0$. It is

$$q^{(1)}(t) = \varepsilon\left(\frac{\sinh^2 t - t^2}{2\cosh t}\right) + \frac{h}{4}\left(\sinh t + \frac{t+1}{\cosh t}\right). \qquad (4.51)$$

Calculating the change of the energy we obtain

$$E(t) = h + \varepsilon t - \varepsilon \tanh t. \tag{4.52}$$

In the vicinities of hyperbolic points in the phase space, $q = \pm \pi$, $p = 0$, we can write for asymptotic behavior of trajectory $q(t) \approx q^{(0)}(t) + q^{(1)}(t)$ given by (4.50) and (4.51) when $t \to \mp \infty$:

$$q_{\mp}(t) = \begin{cases} -\pi + 4\exp(t) - \dfrac{1}{8}(h + 2\varepsilon)\exp(-t), & t \to -\infty, \\[2mm] \pi - 4\exp(t) - \dfrac{1}{8}(h - 2\varepsilon)\exp(-t), & t \to +\infty. \end{cases} \tag{4.53}$$

In the neighborhoods of the X-points ($q = \pm\pi$, $p = 0$) in the phase plane we linearize equation (4.49) and find

$$q_{\mp}(t) = \mp\pi + A_{\pm}\exp(t) + B_{\pm}\exp(-t). \tag{4.54}$$

We have already calculated the adiabatic invariant for nonlinear pendulum. It is given by expression (3.207) which we rewrite in the form

$$J^{(0)}(h, A(t)) = \frac{8}{\pi} \begin{cases} (2h)^{1/2}\left(E(\kappa) - (1 - \kappa^2)K(\kappa)\right), & \kappa < 1 \\[2mm] (A(t))^{1/2}\dfrac{\kappa}{2}E\left(\dfrac{1}{\kappa}\right), & \kappa > 1 \end{cases}, \tag{4.55}$$

with $\kappa = (h/2A)^{1/2}$. Near the separatrix we have

$$J^{(0)} \approx \frac{4}{\pi}\left(2 + (\kappa - 1)\ln\frac{8}{|\kappa - 1|} + (\kappa - 1) + \varepsilon t\right). \tag{4.56}$$

The particle crosses the line $x = -\pi$ at time $t_- = \ln(64/(h + 2\varepsilon))$ with the parameter $\kappa_- = 1 + h/4 + \varepsilon/2$. It reaches the vicinity of the point $x = \pi$ at time $t_+ = \ln(64/(-h + 2\varepsilon))$ with the parameter $\kappa_+ = 1 + h/4 - \varepsilon/2$. During the first step of the particle motion near the separatrix the adiabatic invariant change is

$$\Delta J^{(1)} \approx \frac{1}{\pi}\left(h\ln\left(\frac{2\varepsilon + h}{2\varepsilon - h}\right) - \varepsilon\right). \tag{4.57}$$

Calculating the adiabatic invariant changes step by step we obtain

$$\Delta J \approx \frac{2\varepsilon}{\pi}\left(\beta\ln\left(\frac{1 + \beta}{1 - \beta}\right) - 2 \right.$$

$$\left. - \sum_{n=1}^{\infty}\left(\beta\ln\left(\frac{4n^2 - (1 + \beta)^2}{4n^2 - (1 - \beta)^2}\right) + 2n\ln\left(\frac{(2n + 1)^2 - \beta^2}{(2n - 1)^2 - \beta^2}\right) - 4\right)\right). \tag{4.58}$$

Here we introduce $\beta = h_0/2\varepsilon$ with $h_0 = \varepsilon\sin(q_s/2)/2$ and q_s being the coordinate of the separatrix crossing. The expression (4.58) can be rewritten as

$$\frac{\pi\Delta J}{4\varepsilon} = \lim_{n\to\infty}\frac{\left(\left(n + \dfrac{1}{2}\right)^2 - \dfrac{\beta^2}{4}\right)\pi\exp(2n)}{\Gamma\left(n + \dfrac{1}{2} + \dfrac{\beta}{2}\right)\Gamma\left(n + \dfrac{1}{2} - \dfrac{\beta}{2}\right)\cos\left(\dfrac{\pi\beta}{2}\right)} - 1$$

$$= \ln\left(2\cos\left(\frac{\pi h_0}{\varepsilon}\right)\right). \tag{4.59}$$

Here $\Gamma(x)$ is the gamma function.

Above we have assumed that $\varepsilon \neq h/2$, that is the trajectory is not the new branch of the separatrix. For $|h| = 2\varepsilon$ the trajectory asymptotically tends to (or from) the saddle point as $t \to \pm\infty$. However the number of trajectories with $\varepsilon \approx h/2$ is exponentially small if one assumes the distribution of particles on the phase coordinates to be uniform.

4.4 The Mel'nikov Theory of the Motion near Separatrix

We consider the dynamic system which is described by equations

$$\dot{\mathbf{x}} = \mathbf{f}^0(\mathbf{x}) + \varepsilon \mathbf{f}^1(\mathbf{x}, t), \tag{4.60}$$

with $\varepsilon \ll 1$ and $\mathbf{x} = (\mathbf{q}, \mathbf{p})$, $f_i^0 = J_{ij}\partial_j \mathcal{H}$, where a matrix J_{ij} has been defined in above. We assume the unperturbed motion $\mathbf{x}^{(0)}(t)$ which obeys equation (4.60) with $\mathbf{f}^1 = 0$ to have the separatrix trajectory with the saddle point at $\mathbf{x}_{\text{saddle}}$: On the separatrix we have

$$\lim_{t \to \pm\infty} \mathbf{x}_s^{(0)}(t) = \mathbf{x}_{\text{saddle}}. \tag{4.61}$$

The perturbations change the position of the saddle point on $\delta\mathbf{x}_{\text{saddle}}$ and the form of the separatrix. Instead of a loop we have two branches of the separatrix: $x_+(t)$ and $x_-(t)$. From equation (4.60) it follows that

$$\dot{x}_i^{(1)} = w_{ij}(\mathbf{x}^{(0)})x_j^{(1)} + \varepsilon f_i^1(\mathbf{x}^{(0)}, t), \tag{4.62}$$

where $w_{ij} = \partial f_i^0/\partial x_j = \dot{M}_{ik}M_{kj}^{-1}$.

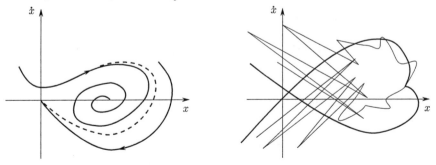

Fig. 4.4. Change of the separatrix under perturbations.

Depending on the structure of perturbations the separatrix initially enclosing the center-type critical point in the phase plane bifurcates into either the stable (an attractor) or the unstable (a repeller) critical point or transforms into the homoclinic tangle configuration shown in Fig. 4.4.

Following Mel'nikov we find the threshold for the *homoclinic tangle* structure to appear. We need to calculate the distance between two new branches $\mathbf{x}_+$ and $\mathbf{x}_-$ of the separatrix. The distance is a vector $\mathbf{d}(s) = \mathbf{x}_+(s, s_0) - \mathbf{x}_-(s, s_0)$, which has a name of Mel'nikov's vector. Here s is a coordinate along the separatrix:

$$ds = |\nabla \mathcal{H}| dt \qquad (4.63)$$

and s_0 is the initial position. The vector $\mathbf{d}(s)$ is shown in Fig. 4.5.

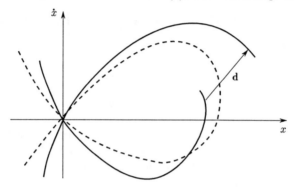

Fig. 4.5. Mel'nikov's vector $\mathbf{d}$.

The scalar product of the Mel'nikov vector, $\mathbf{d}$, and $\mathbf{N}$, a vector of the normal to the unperturbed separatrix at the point s_0, is

$$D(s, s_0) = (\mathbf{d}(s, s_0) \cdot \mathbf{N}(s, s_0)). \qquad (4.64)$$

Since unperturbed separatrix is a surface of constant value of the Hamiltonian, $\mathcal{H}(\mathbf{p}, \mathbf{q})$, the normal to the unperturbed separatrix is

$$\mathbf{N} = \nabla \mathcal{H}(\mathbf{p}^{(0)}(s), \mathbf{q}^{(0)}(s)). \qquad (4.65)$$

Now we introduce functions $D_\mp(s, s_0) = (\mathbf{x}_\mp \cdot \mathbf{N})$ and calculate a time derivative

$$\dot{D}_\mp(s, s_0) = (\dot{\mathbf{x}}_\mp \cdot \mathbf{N} + \mathbf{x}_\mp \cdot \dot{\mathbf{N}}) = (w_{ij}\partial_j\mathcal{H} + \partial_i \dot{H})x_{\mp i} + \varepsilon f_i^1(t)\partial_i\mathcal{H}$$

$$= (w_{ji}\partial_j\mathcal{H} + w_{ik}J_{kl}\partial_l\mathcal{H})x_{\mp i} + \varepsilon f_i^1(t)\partial_i\mathcal{H} = \varepsilon f_i^1(t)\partial_i\mathcal{H}(t - t_0). \qquad (4.66)$$

Using the boundary conditions, $\lim_{t \to \pm\infty} \mathbf{x}_\pm = \delta\mathbf{x}_{\text{saddle}}$, we obtain

$$D(s_0) \equiv D(s_0, s_0) = -\varepsilon \int_{-\infty}^{+\infty} f_i^1(t)\partial_i\mathcal{H}(t - t_0)dt$$

$$= -\varepsilon \oint f_i^1(s)\partial_i\mathcal{H}(s - s_0)\frac{ds}{|\partial_j\mathcal{H}|} = -\varepsilon \int_{-\infty}^{+\infty} f_i^1(t)J_{ij}f_j^0(t - t_0)dt. \qquad (4.67)$$

A condition for the *homoclinic tangle* structure to appear corresponds to the change of the sign of the function $D(s_0)$.

Calculating the vector $\mathbf{d}$ with the help of relationship (7.35) with

$$\varepsilon f(x, \dot{x}, t) = -2\varepsilon\nu\dot{x} + \varepsilon f \sin \Omega t, \qquad (4.68)$$

we obtain for $t \to \infty$

$$D(t_0) = \varepsilon \left(\sqrt{2} f \sin \Omega t_0 \int_{-\infty}^{+\infty} dt' \frac{\sinh t'}{\cosh^2 t'} \sin \Omega t' + 2\nu \int_{-\infty}^{+\infty} dt' \frac{\tanh^2 t'}{\cosh^2 t'} \right). \qquad (4.69)$$

Evaluating the integrals in this expression we find

$$D(t_0) = \varepsilon \left(\sqrt{2}\pi f \frac{\sin \Omega t_0}{\cosh (\pi\Omega/2)} + \frac{4\nu}{3} \right). \qquad (4.70)$$

We see that if the dissipation is high enough, $D(t_0)$ does not change its sign. It changes its sign when

$$\nu < \nu_c = \frac{3\pi f}{2\sqrt{2}\cosh (\pi\Omega/2)}. \qquad (4.71)$$

This is a threshold for appearance of the homoclinic tangle and then for chaos regime to appear.

4.5 Change of Adiabatic Invariant During the Separatrix Birth

Let us consider the particle motion near the critical point where the separatrix birth occurs. The time dependent Hamiltonian has a form

$$\mathcal{H}(p, q, \varepsilon t) = \frac{p^2}{2} - \varepsilon t \frac{q^2}{2} + \frac{q^3}{3} + \frac{q^4}{4} = h. \qquad (4.72)$$

The separatrix birth ($h = 0$) takes place for $\varepsilon t = 0$. If $t < 0$, the frozen time dependence of the Hamiltonian corresponds to one region in a phase space containing closed trajectories, for $t > 0$ we have two regions with particle trajectories with $h < 0$. The potential energy has one minimum for $t < 0$ and two minima for $t > 0$.

The equation of motion,

$$\ddot{q} - \varepsilon t q + q^2 + q^3 = 0, \qquad (4.73)$$

in zeroth order of ε has a solution corresponding to the motion on the separatrix when $h = 0$:

$$q^{(0)} = -\frac{12}{2t^2 + 9}, \qquad \dot{q}^{(0)} = \frac{48t}{(2t^2 + 9)^2}. \qquad (4.74)$$

Near the critical point in the phase space $p = 0$, $q = 0$, the trajectory has the shape of a cusp with $p \propto q^{3/2}$.

In the first order of ε from equation (4.73) we have for $q^{(1)}(t)$ equation

$$\ddot{q}^{(1)} + 2q^{(0)}q^{(1)} + 3\left(q^{(0)}\right)^2 q^{(1)} = \varepsilon t q^{(0)} \tag{4.75}$$

with a solution

$$q^{(1)} = \dot{q}^{(0)} \int^t \frac{d\tilde{t}}{\left(\dot{q}^{(0)}\right)^2}\left(\frac{4}{9}h + \varepsilon \int^{\tilde{t}} \tilde{t}\dot{q}^{(0)}q^{(0)}d\tilde{t}\right). \tag{4.76}$$

When $t \to \pm\infty$, the asymptotic dependence of zeroth order solution is

$$q^{(0)} \propto t^{-2}, \quad \dot{q}^{(0)} \propto t^3. \tag{4.77}$$

As a result we have

$$q_\mp(t) = q^{(0)}(t) + q^{(1)}(t) \approx -\frac{6}{t^2} + \frac{1}{84}\left(h \mp \frac{\pi\varepsilon}{3\sqrt{2}}\right)t^4 \tag{4.78}$$

as $t \to \mp\infty$. The trajectory crosses the ordinate axis in a phase space at the time values t_- and t_+ respectively. These time values are found from the requirement for the solution to vanish $(q(t_\mp) = 0)$. From equation (4.78) we obtain

$$t_\mp = \mp \left(\frac{1512 \cdot \sqrt{2}}{3\sqrt{2}h \mp \pi\varepsilon}\right)^{1/6}. \tag{4.79}$$

For $h \to \pm\pi\varepsilon/(3\sqrt{2})$ these values tend to infinity that corresponds to the motion along the incoming and outgoing vicinity of the critical point separatrix branches. As we have noted above, under the action of perturbations the separatrix splitting occurs into two branches.

Substitution of the expressions (4.76)–(4.79) into the expression for the change of adiabatic invariant during one step of motion,

$$\Delta J = \frac{2}{\pi}\int_{t_-}^{t_+} q^{(1)}\dot{q}^{(0)}dt, \tag{4.80}$$

leads to the estimation

$$\Delta J = \frac{8}{7\pi}(504)^{1/6}\left(\left(h - \frac{\pi\varepsilon}{3\sqrt{2}}\right)^{5/6} - \left(h + \frac{\pi\varepsilon}{3\sqrt{2}}\right)^{5/6}\right). \tag{4.81}$$

This formula is in agreement with the value estimated by formula (4.18) for $n = 3$.

4.5.1 Condition of Stochastic Regime Appearance in the Course of the Separatrix Birth

Let us discuss the problem of appearance of a stochastic regime of the motion near the separatrix given by equations (4.74). We calculated the Mel'nikov distance which is the measure of the separatrix splitting (4.67). We write the equation of motion corresponding to the perturbed separatrix as

$$\ddot{q} + q^2 + q^3 = -\varepsilon\delta\dot{q} + \varepsilon\gamma\cos\omega t. \tag{4.82}$$

This equation describes the periodic motion in the vicinity of the separatrix (see equation (4.73) for $t = 0$), perturbed by small damping $\varepsilon\delta$ and the action of the periodic force with the amplitude $\varepsilon\gamma$. Calculation of the vectors $\mathbf{N}$ and $\mathbf{d}$ gives for the value $D(t, t_0)$ in correspondence with equation (4.67)

$$D(t_0) = -48\gamma\sin\omega t_0 \int\limits_{-\infty}^{+\infty} d\tau\frac{\tau}{(2\tau^2 + 9)^2}\sin\omega\tau + +576\delta \int\limits_{-\infty}^{+\infty}\frac{\tau^2}{(2\tau^2 + 9)^4}d\tau$$

$$= -12\pi\gamma\sin\omega t_0 \exp\left(-\frac{3}{2^{1/2}}\omega\right) + \frac{2^{5/2}}{27}\pi\delta. \tag{4.83}$$

The stochastic motion near the separatrix occurs when the perturbed trajectories are crossing, i.e. when the function $D(t_0)$ changes its sign. From (4.83) it follows that the condition of this is

$$\delta < \delta_c = \frac{81}{2^{1/2}}\gamma\exp\left(-\frac{3}{2^{1/2}}\omega\right). \tag{4.84}$$

4.6 Nonlinear Resonances, their Overlap, Standard Mapping

Above we have discussed motion of the system with two degrees of freedom. We have seen that in this case a system could have the internal resonance, the condition of which we might write as $n\omega_1 + m\omega_2 = 0$. Formally it corresponds to the resonance response of one of the oscillators on periodic force that appears due to nonlinear interaction between different degrees of freedom. Now we consider the motion of a particle in the case when the Hamiltonian is periodic with the frequency Ω being the function of time. We assume that time dependence of the Hamiltonian is given by

$$\mathcal{H} = \mathcal{H}_0(J) + \varepsilon V(J, \theta, t) = \mathcal{H}_0(J) + \varepsilon \sum_{n,m=-\infty}^{+\infty} V_{n,m}(J)\exp(i(n\theta - m\Omega t)) \tag{4.85}$$

with $V_{-n,-m}(J) = V_{n,m}^*(J)$. We use the action-angle coordinates. The parameter ε is assumed to be small. From equation (4.85) we find that the resonance condition is

$$n\omega(J) - m\Omega = 0, \tag{4.86}$$

where

$$w(J) = \frac{\partial \mathcal{H}_0}{\partial J} \tag{4.87}$$

is the frequency of unperturbed oscillations. Equation (4.86) is the equation for the action J_0 for which the resonance condition (4.85) holds.

Similarly to the case considered in Lecture 4 we retain only the resonant term in the right-hand side of equation (4.86) and obtain

$$\dot{J} = -\frac{\partial \mathcal{H}_0}{\partial \theta} \approx -2\varepsilon n V_0 \sin \psi. \tag{4.88}$$

Here we use a notation for the amplitude of perturbation: $V_0 = V_{n,m}(J_0)$ and introduce the phase

$$\psi = n\theta - m\Omega t. \tag{4.89}$$

Differentiating ψ with respect to time two times and using (4.87) and (4.88) we find an equation for the phase at nonlinear resonance:

$$\ddot{\psi} + \omega_0^2 \sin \psi = 0. \tag{4.90}$$

This is the equation of nonlinear pendulum which we have studied above in details. The frequency ω_0 is the frequency of small amplitude oscillations or it is also called the "frequency width" of the resonance $\Delta\omega$. It is

$$\omega_0 = \left(\varepsilon n^2 V_0 \left| \frac{\partial^2 \mathcal{H}_0}{\partial J^2} \right| \right)^{1/2}. \tag{4.91}$$

Correspondingly, the width of the resonance in the action is

$$\Delta J = \frac{\Delta\omega}{|\partial^2 \mathcal{H}_0/\partial J^2|} = \left(\frac{\varepsilon n^2 V_0}{|\partial^2 \mathcal{H}_0/\partial J^2|} \right)^{1/2}. \tag{4.92}$$

In general, there can be several resonances in the Hamiltonian system with the Hamiltonian (4.85). We say δJ is a distance between the resonant values $J_{n,m}$, then the distance between resonances in the frequency scale is given by

$$\delta\omega = \left| \frac{\partial\omega}{\partial J} \right| \delta J. \tag{4.93}$$

The basic parameter determining the evolution of the system describes the degree of overlap of different resonances:

$$\bar{K} = \frac{\Delta\omega}{\delta\omega}. \tag{4.94}$$

At $\bar{K} \ll 1$ the motion is regular (periodic), while at $\bar{K} \geq 1$ the dynamics of the system becomes stochastic (B.V. Chirikov, 1959).

Now we consider a different approach. We assume that $n = \pm 1$ in (4.85). The Hamiltonian (4.85) may be written as

$$\mathcal{H} = \mathcal{H}_0(J) - 2\varepsilon T V_0 \cos\theta \sum_{m=-\infty}^{+\infty} \delta(t - mT) \qquad (4.95)$$

with $V_0 = V_{\pm 1,m}(J)$ and $T = 2\pi/\Omega$. This approximation corresponds to a non-linear oscillator which is acted upon periodically by δ−function jolts. Assuming for simplicity that V_0 does not depend on J we denote by (J_n, θ_n) the values of action and angle variables at time $t_n = nT$. The equation of motion then can be written as the mapping

$$\begin{aligned} J_{n+1} &= J_n - 2\varepsilon T V_0 \sin\theta_n, \qquad (4.96) \\ \theta_{n+1} &= \theta_n + T\omega(J_{n+1}). \end{aligned}$$

The parameter related to $\bar{K}$ in this case is

$$K = 2\varepsilon T^2 V_0 \left|\frac{\partial\omega}{\partial J}\right| \cos\theta. \qquad (4.97)$$

At $K \ll 1$ the motion is regular , while at $K \geq 1$ it becomes stochastic everywhere except in "stability islands". There is a relationship between these two parameters: $K \sim \bar{K}^2$.

4.7   Strange Attractor. The Lorenz System.

Now we consider the well known Lorenz system, which is a system of three ordinary differential equations of the first order. It describes nonlinear regime of the convective instability in a slab of the fluid (see Fig. 4.6). The slab of incompressible fluid is assumed to have a width l along the z−axis and to be in the gravitation field, g. The temperature at the bottom, $T + \delta T$, is assumed to be larger than the temperature at the top of the slab, T.

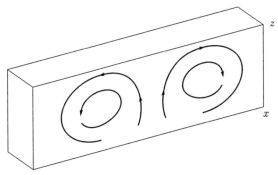

Fig. 4.6. Convection in a fluid slab.

We shall discuss the equations of motion of incompressible fluid below. In the considered case they can be written as

$$\partial_t \omega + \{\psi, \omega\} = \nu \Delta \omega - g a \partial_x \vartheta, \tag{4.98}$$

$$\partial_t \vartheta + \{\psi, \vartheta\} = \kappa \Delta \vartheta + \delta T \partial_x \psi. \tag{4.99}$$

Here Ψ is the stream function, i.e. the fluid velocity is equal to $\mathbf{v} = \nabla(\psi \mathbf{e}_y)$, $\omega = \nabla \times \mathbf{v} = -\Delta \psi$, ϑ is the temperature, ν and κ are the coefficients of viscosity and thermal conductivity, respectively, and $\{f, h\} = \partial_x f \partial_z h - \partial_z f \partial_x h$ are the Poisson brackets. We seek a solution of the linearized problem in the form

$$\psi = \psi(t) \sin\left(\frac{\pi k x}{l}\right) \sin\left(\frac{\pi z}{l}\right), \text{ and } \vartheta = \vartheta(t) \cos\left(\frac{\pi k x}{l}\right) \sin\left(\frac{\pi z}{l}\right), \tag{4.100}$$

where k is the dimensionless wavenumber in the x–direction.

The problem is characterized by the dimensionless parameter

$$R_a = g a l^3 \frac{\delta T}{\kappa \nu}, \tag{4.101}$$

the Rayleigh number. In a linear approximation it can be shown that the system is unstable when the Rayleigh number is larger than some critical value which is equal to $R_c = \pi^4 (1 + k^2)^3$. The maximum value parameter $R_c = 27\pi^4/4$ reaches for the wavenumber $k = 1/\sqrt{2}$.

We describe the nonlinear regime of the fluid convection with the functions

$$\psi = C_1 \sqrt{2} X(t) \sin\left(\frac{\pi k x}{l}\right) \sin\left(\frac{\pi z}{l}\right),$$

$$\vartheta = C_2 \left(\sqrt{2} Y(y) \left(\frac{\pi k x}{l}\right) \sin\left(\frac{\pi z}{l}\right) - Z(t) \sin\left(\frac{\pi z}{l}\right) \right). \tag{4.102}$$

Substituting ψ and ϑ (4.102) into equations (4.98,4.99) we obtain the Lorenz equations

$$\dot{X} = -\sigma(X - Y), \tag{4.103}$$

$$\dot{Y} = rX - Y - XZ, \tag{4.104}$$

$$\dot{Z} = -bZ + XY. \tag{4.105}$$

Here the time variable is normalized as $\tau = \pi^2(1 - k^2)\kappa t/l$; dimensionless parameters are $\sigma = R_a/R_c$, $b = 4/(1 + k^2)$.

The Lorenz system describes a flux in the space (X, Y, Z) with the velocity $\mathbf{V} = (\dot{X}, \dot{Y}, \dot{Z})$. Calculating $\operatorname{div} \mathbf{V}$, which is equal to $\sigma + b + 1$, we find that the element of the volume in the space (X, Y, Z) decreases exponentially $\Gamma(t) = \Gamma(0) \exp(-(\sigma + b + 1)t)$. It means that the trajectories of system (4.103)-(4.105) are attracted by the region with the zero measure in the space (X, Y, Z). The attractor can be either regular or strange (solution of the Cauchy problem shows very complicated evolution, which corresponds to stochastic behavior). System (4.103)-(4.105) always has an equilibrium solution $(X_0 = Y_0 = Z_0 = 0)$ and can have two more equilibrium points, $X_\pm = Y_\pm = \pm\sqrt{b(r-1)}$, $Z_\pm = (r-1)$, either stable or unstable, depending on the parameters σ, r, b. Linearizing system (4.103-4.105) around equilibrium $X_0 = Y_0 = Z_0 = 0$, and assuming dependence of perturbations on time to be $\sim \exp(\gamma t)$, we find the dispersion equation for the growth rate γ:

$$(\gamma + b)(\gamma^2 + (\sigma + 1)\gamma + \sigma(1 - r)) = 0. \tag{4.106}$$

We see that for $0 < r < 1$ the equilibrium (X_0, Y_0, Z_0) is stable. Similarly we may investigate stability of the equilibria $(X_\pm, Y_\pm, Z_\pm)$.

For $\sigma = 10, r = 28, b = 8/3$ the attractor is strange. The chaotic trajectory in 3D space (X, Y, Z) is shown in Fig. 4.7.

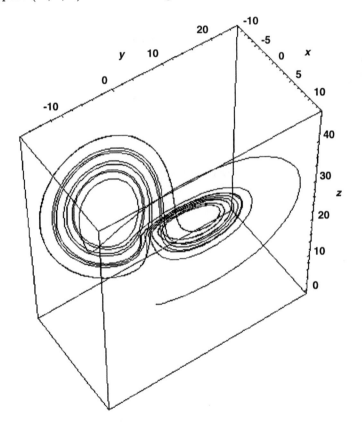

Fig. 4.7. Strange attractor

Introducing a new variable $q = z - x^2/2\sigma$, we transform the Lorenz system to

$$\ddot{x} + (1 + \sigma)\dot{x} - \sigma(1 - r - q)x + x^3/2 = 0, \tag{4.107}$$

$$\dot{q} + bq = (1 - b/2\sigma)x^2. \tag{4.108}$$

Equation (4.107) describes the anharmonic oscillator with damping and frequency that varies via dependence of the function $q(t)$ on time. The latter, in turn, depends on the amplitude of oscillations via equation (4.108).

For frozen dependence of the function q on time, equation (4.107) describes motion of the particle in the potential which has two minima for $q > 1 - r$ and one minimum for $q < 1 - r$. Since by virtue of equation (4.108) q depends on time, the particle moves in the potential well, where form and depth vary in time. In the $(\dot{x}, x)$ phase, the plane position and form of the separatrix change (the separatrix appears and disappears) and the trajectory is crossing the separatrix. In Fig. 4.8 we show $(\dot{x} \equiv v, x)$ phase plane (a), and the $q(t)$ variable.

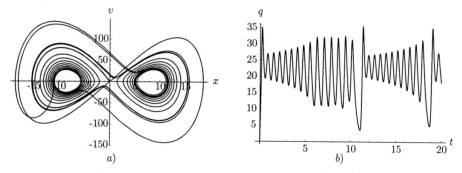

Fig. 4.8. Strange attractor in coordinates $\dot{x}, x$, (a), and q versus time (b).

◇

Part II

Waves

Part II

Waves

Chapter 5

Linear Waves in Dispersive Media

Linear wave propagation in dispersive media. Phase and group velocities. The stationary phase method to describe the wave packet propagation. Long time asymptotics. Propagation of electromagnetic wave packets in plasmas. Geometric optics. Wave propagation in nonhomogeneous media. Nonadiabatic regimes of wave propagation. Wave behavior near the focus.

$$***$$

In this section we discuss types of equations that describe wave propagation in different media.

In the simplest case the wave is given by the dependence

$$u(x,t) = f(x \pm ct). \qquad (5.1)$$

Asymptotically it corresponds to noninteracting waves that propagate in positive and negative directions along the x-axis as it is shown in Fig. 5.1.

The wave propagating to the left depends on the variable $x + ct$, meanwhile for the one that propagates to the right $x - ct = $ const. This means that the dependences satisfy the equation $(\partial_t \mp c\partial_x) f = 0$. These equations can be obtained by factorization of the wave operator $\partial_{tt} - c^2\partial_{xx}$ as follows:

$$\left(\partial_{tt} - c^2\partial_{xx}\right) u = \left(\partial_t - c\partial_x\right)\left(\partial_t + c\partial_x\right) u = 0. \qquad (5.2)$$

We see that two independent waves propagating in negative and positive directions along the $x-$axis are described by more simple equations

$$\left(\partial_t - c\partial_x\right) u = 0 \qquad \text{and} \qquad \left(\partial_t + c\partial_x\right) u = 0. \qquad (5.3)$$

In a general case a function that describes linear waves in uniform, time independent media can be written in the form

$$u(x,t) = U_{k\omega} \exp\left(-i(\omega t - \mathbf{k} \cdot \mathbf{x})\right). \qquad (5.4)$$

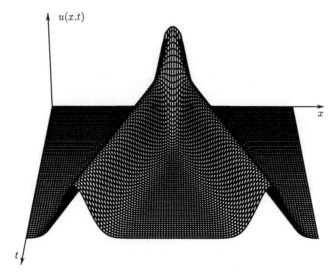

Fig. 5.1. Waves.

In the absence of any dissipation both ω and $\mathbf{k}$ are real. That follows from our assumption that the media is uniform and time independent. In waveguides there are "nonuniform" waves with $\mathbf{k} = \mathbf{k}' + ik''$. However, expression (5.4) does not correspond to any wave until no relationship between the frequency ω and the wavevector $\mathbf{k}$ is imposed. Such a relationship is given by the dispersion equation

$$G(\omega, \mathbf{k}) = 0, \tag{5.5}$$

or, if the equation is resolved with respect to ω, we have

$$\omega = W(\mathbf{k}). \tag{5.6}$$

In the one-dimensional case the frequency is a function of the wavenumber $k : \omega = W(k)$. As a result from expression (5.4) we have

$$u(x, t) = U \left\{ \begin{array}{c} \sin \\ \cos \end{array} \right\} (\omega t - kx) \equiv U \left\{ \begin{array}{c} \sin \\ \cos \end{array} \right\} \theta, \tag{5.7}$$

where the phase θ is

$$\theta = \omega t - kx = \omega \left(t - \frac{k}{\omega} x \right) \equiv k \left(\frac{\omega}{k} t - x \right) \equiv -k(x - v_{\mathrm{ph}} t). \tag{5.8}$$

The point of a constant phase propagates with the phase velocity

$$v_{\mathrm{ph}} = \frac{\omega}{k} = \frac{W(k)}{k}. \tag{5.9}$$

It is a function of the wavenumber.

Now we consider the linear superposition of two sinusoidal waves with equal amplitudes, U, and different frequencies $\omega \pm \Delta \omega$ and wavenumbers $k \pm \Delta k$ for the first and the second wave, respectively.

$$u(x,t) = U(\cos((\omega + \Delta\omega)t - (k + \Delta k)x) + \cos((\omega - \Delta\omega)t - (k - \Delta k)x))$$

$$= U\cos(\omega t - kx)\cos(\Delta\omega t - \Delta kx). \tag{5.10}$$

This expression describes the beat wave.

If $\Delta\omega \ll \omega$ the points of a constant phase propagate with the phase velocity, $v_{ph} = \omega/k$, meanwhile the points of the amplitude maximum propagate with the group velocity, it is $v_g = \Delta\omega/\Delta k$. When $\Delta k \to 0$ we obtain $v_g = \partial_k\omega$, or

$$v_g = \partial_k W. \tag{5.11}$$

If $W' = W/k$, that is the phase velocity is equal to the group velocity, $v_{ph} = v_g$, when $W'' = 0$, and W is a linear function of k, $W = v_{ph}k$, the medium is nondispersive. The wave packet propagates there without any change of its shape.

When $W'' \neq 0$, $v_{ph} = W/k \neq v_g = W'$.

An electromagnetic wave in a plasma in one dimensional approximation is described by equation

$$\partial_{tt}u - c^2\partial_{xx}u + \omega_{pe}^2 u = 0, \tag{5.12}$$

that provides an example of the wave with a dispersion. Here the Langmuir wave frequency is $\omega_{pe} = (4\pi ne^2/m)^{1/2}$. The dispersion equation is

$$\omega = \pm(k^2c^2 + \omega_{pe}^2)^{1/2}. \tag{5.13}$$

The phase and group velocities are equal to

$$v_{ph} = (k^2c^2 + \omega_{pe}^2)^{1/2}/k, \quad \text{and} \quad v_g = kc^2/(k^2c^2 + \omega_{pe}^2)^{1/2}, \tag{5.14}$$

respectively. We see that $v_{ph} \neq v_g$.

Electromagnetic waves in a plasma in a three-dimensional approximation are described by the equation

$$\partial_{tt}u - c^2\Delta u + \omega_{pe}^2 u = 0. \tag{5.15}$$

Here the Laplace operator is

$$\Delta = \partial_{xx} + \partial_{yy} + \partial_{zz}. \tag{5.16}$$

The linear KdV (Korteweg & de Vries) equation

$$\partial_t u + c_0\partial_x u + \beta\partial_{xxx}u = 0, \tag{5.17}$$

with the dispersion equation

$$\omega = kc_0 + \beta k^3 \tag{5.18}$$

describes parabolic dependence of the phase and group velocity on the wavenumber:

$$v_{ph} = c_0 + \beta k^2, \quad \text{and} \quad v_g = c_0 + 3\beta k^2. \tag{5.19}$$

We see that in this case the phase and group velocities are not equal to each other, $v_{\mathrm{ph}} \neq v_{\mathrm{g}}$.

The 3D version of the KdV equation is known as the Kadomtsev-Petviashvili (K-P) equation. In the linear case it can be written as

$$\partial_x(\partial_t + c_0\partial_x + \beta\partial_{xxx})u = \frac{c_0}{2}\Delta_\perp u, \tag{5.20}$$

where

$$\Delta_\perp = \partial_{yy} + \partial_{zz}. \tag{5.21}$$

Both the KdV and K-P equations correspond to the limit of a weak dispersion, that is $\beta k^3 \ll k c_0$. In this limit in 3D case we write the phase velocity as

$$v_{\mathrm{ph}} = c_0 + \beta k^2 \equiv c_0 + \beta(k_x^2 + k_y^2 + k_z^2). \tag{5.22}$$

Then we have for the wave frequency

$$w = k \cdot v_{\mathrm{ph}} = k \cdot (c_0 + \beta k^2) \equiv (k_x^2 + k_y^2 + k_z^2)^{1/2}(c_0 + \beta(k_x^2 + k_y^2 + k_z^2))$$

$$\approx k_x c_0 + \beta k_x^3 + c_0\frac{k_y^2 + k_z^2}{2k_x}. \tag{5.23}$$

The terms $\beta k_x^3 + c_0(k_y^2 + k_z^2)/2k_x$ are of the same order and small compared with $k_x c_0$. Here we have represented the absolute value of the wavevector $k = (k_x^2 + k_\perp^2)^{1/2}$ as $k \approx k_x + k_\perp^2/2k_x$ for $k_x \gg k_\perp$. Then we rewrite equation (5.23) in the form

$$k_x(w - k_x c_0 - \beta k_x^3)u = -\frac{c_0}{2}(k_y^2 + k_z^2)u. \tag{5.24}$$

Invoking relationships $w \leftrightarrow i\partial_t$, $k_x \leftrightarrow -i\partial_x$, and $k_\perp^2 \leftrightarrow -\Delta_\perp$, we obtain equation (5.20).

The most general form of the linear wave equation in a dispersive media is

$$\partial_t u + \int_{-\infty}^{+\infty} K(x - \xi)d\xi = 0. \tag{5.25}$$

This is the Witham equation. The Fourier transforms of the Whitham equation with respect to the time and space coordinate give

$$G(\omega, k) \equiv \left(w + i\int_{-\infty}^{+\infty} K(\xi)\exp(-ik\xi)d\xi\right) = 0. \tag{5.26}$$

That is the dispersion equation $\omega = W(k)$ can be written as

$$\omega = W(k) = -i\int_{-\infty}^{+\infty} K(\xi)\exp(-ik\xi)d\xi. \tag{5.27}$$

Exercise. Show that the dispersion equation for the Schrödinger equation

$$i\partial_t u + \Delta u = 0 \tag{5.28}$$

is $\omega = k^2$.

◇

5.1 Asymptotic Behavior of the Wave Packet in Dispersive Media

A generic expression that describes the wave dependence on time and coordinate can be written in 1D case as

$$u(x,t) = \int_{-\infty}^{+\infty} F(k) \exp(i(kx - W(k)t) \frac{dk}{2\pi}. \tag{5.29}$$

What does this expression come from?

The Cauchy Problem for the Linear Differential Equation in Partial Derivatives

To describe the evolution of a small amplitude perturbation in continuous media we must solve the linear differential equation in partial derivatives

$$G_{jk}(\partial_l, \partial_t) u_k(x_l, t) = 0 \tag{5.30}$$

with initial conditions, $u_k(x_l, 0), \partial_t u_k(x_l, 0), \ldots$. Here $\hat{\mathbf{G}}(\partial_l, \partial_t)$ is a linear differential operator, $\partial_t = \partial/\partial t$, $\partial_l = \partial/\partial x_l$. The Laplace-Fourier transforms give for the function

$$U_k(k_l, \omega) = \int_{-\infty}^{+\infty} dx \int_{-\infty}^{+\infty} dy \int_{-\infty}^{+\infty} dz \int_{0}^{+\infty} dt \exp(i(\omega t - k_l x_l)\, u_k(x_l, t) \tag{5.31}$$

the equation

$$G_{jk}(ik_l, -i\omega)U_k(k_l, \omega) = g_j(k_l, \omega). \tag{5.32}$$

In the right-hand side a vector $g_j(k_l, \omega)$ is determined by initial conditions. The solution for equation (5.32) is

$$U_k(k_l, \omega) = \frac{S_{kj}(ik_l, -i\omega)g_j(k_l, \omega)}{\det(G_{jk}(ik_l, -i\omega))}, \tag{5.33}$$

where $S_{kj}(ik_l, -i\omega)$ is a cofactor of the matrix $G_{jk}(ik_l, -i\omega)$.

Calculating the inverse Laplace transform of expression (5.33) with respect to variable ω we find the function

$$F_k(k_l, t) = \frac{1}{2\pi} \oint_S U_k(k_l, \omega) \exp(-i\omega t) d\omega$$

$$= \frac{1}{2\pi} \oint_S \frac{S_{kj}(ik_l, -i\omega)g_j(k_l, \omega)}{\det(G_{jk}(ik_l, -i\omega))} \exp(-i\omega t) d\omega. \tag{5.34}$$

Here S is a contour in the complex variable $\omega = \omega' + i\omega''$ chosen to fulfill the causality requirements. Asymptotically, when $t \to \infty$, the value of the integral is determined by the poles of the integrand, that is by zeros of the denominator: $\det(G_{jk}(ik_l, -i\omega)) = 0$. This expression is the dispersion equation. If it can be rewritten as $\omega^\alpha = W^\alpha(k_l)$, where α corresponds to a particular mode, we obtain

$$F_k(k_l, t) \approx \sum_\alpha F_k^\alpha(k_l) \exp(-iW^\alpha(k_l)t). \qquad (5.35)$$

Here $F_k^\alpha(k_l)$ and $W^\alpha(k_l)$ are the amplitude and frequency of the mode α. Equation (5.29) is an inverse Fourier transform with respect to the variable k, written in 1D case for a particular mode.

5.2   Electromagnetic Waves

Now we illustrate the procedure to find the dispersion equation by the study of electromagnetic wave propagation in dispersive media. As it is well known, the electromagnetic waves are described by a set of the Maxwell equations:

$$\text{curl } \mathbf{B} = \frac{1}{c}\partial_t \mathbf{D}, \qquad \text{div } \mathbf{B} = 0, \qquad (5.36)$$

$$\text{curl } \mathbf{E} = -\frac{1}{c}\partial_t \mathbf{B}, \qquad \text{div } \mathbf{D} = 0. \qquad (5.37)$$

The external electric charge and electric current density are assumed to be zero. The magnetic permittivity, μ, is assumed to be an equal unity. In this case the magnetic field, $\mathbf{H}$, is equal to $\mathbf{B}$. Further we shall use the notation $\mathbf{B}$ for the magnetic field. The electric field $\mathbf{E}$, the magnetic field $\mathbf{B}$, and the electric displacement vector $\mathbf{D}$, are related each other via material equation

$$\mathbf{D} = \mathbf{D}(\mathbf{E}, \mathbf{B}). \qquad (5.38)$$

In the limit of small amplitude of the wave we can expand equation (5.38). To the first order in the electric field amplitude we obtain

$$\mathbf{D} = \mathbf{D}_0 + \hat{e}\mathbf{E} + \dots. \qquad (5.39)$$

Here the magnetic field $\mathbf{B}$ was expressed via the electric field. Further we assume also that $\mathbf{D}_0 = 0$.

The action of linear operator $\hat{\varepsilon}$ in the case of time independent and homogeneous media on vector $\mathbf{E}$ can be written as

$$D_i(\mathbf{x}, t) = \int_{c(t-t')<|\mathbf{x}-\mathbf{x}'|} dt' d^3x' \varepsilon_{ij}(t-t', \mathbf{x}-\mathbf{x}') E_j(t', \mathbf{x}'). \qquad (5.40)$$

Here the integration is performed over the region $c(t - t') < |\mathbf{x} - \mathbf{x}'|$ by virtue of causality principle. In the limit $c \to \infty$ formula (5.40) takes the form

$$D_i(\mathbf{x}, t) = \int_0^t dt' \int\int_{-\infty}^{+\infty}\int d^3x' \varepsilon_{ij}(t-t', \mathbf{x}-\mathbf{x}') E_j(t', \mathbf{x}'). \qquad (5.41)$$

The Laplace-Fourier transforms give

$$D_i(\mathbf{k},\omega) = \varepsilon_{ij}(\mathbf{k},\omega)E_j(\mathbf{k},\omega) \tag{5.42}$$

with $\varepsilon_{ij}(\mathbf{k},\omega)$ being the tensor of dielectric permittivity.

From equations (5.36)—(5.37) it follows

$$k_i B_i(\mathbf{k},\omega) = 0, \qquad \frac{\omega}{c}B_i(\mathbf{k},\omega) - \epsilon_{ijk}k_j E_k(\mathbf{k},\omega) = \frac{i}{c}B_i(\mathbf{k},t=0), \tag{5.43}$$

$$k_i D_i(\mathbf{k},\omega) = 0, \qquad \frac{\omega}{c}D_i(\mathbf{k},\omega) + \epsilon_{ijk}k_j B_k(\mathbf{k},\omega) = \frac{i}{c}D_i(\mathbf{k},t=0). \tag{5.44}$$

Here ϵ_{ijk} is the antisymmetric tensor of the third order. These equations can be rewritten in the form

$$G_{ij}(\mathbf{k},\omega)E_j(\mathbf{k},\omega) = \frac{i\omega}{c^2}D_i(\mathbf{k},t=0) + \frac{i}{c}\epsilon_{ijk}k_j B_k(\mathbf{k},t=0) \equiv g_i(\mathbf{k},\omega). \tag{5.45}$$

Solving the equation we find

$$E_i(\mathbf{k},\omega) = \frac{S_{kj}(\mathbf{k},\omega)g_j(\mathbf{k},\omega)}{\det(G_{ij}(\mathbf{k},\omega))}, \tag{5.46}$$

where

$$G_{ij}(\mathbf{k},\omega) = k^2\delta_{ij} - k_i k_j - \frac{\omega^2}{c^2}\varepsilon_{ij}(\mathbf{k},\omega). \tag{5.47}$$

According to formula (5.34) the dispersion equation has the form

$$G(\mathbf{k},\omega) \equiv \det(G_{ij}(\mathbf{k},\omega)) = \det\left(k^2\delta_{ij} - k_i k_j - \frac{\omega^2}{c^2}\varepsilon_{ij}(\mathbf{k},\omega)\right) = 0. \tag{5.48}$$

This is the dispersion equation for the eigenfrequencies of the electromagnetic waves propagating in the media specified by the dielectric tensor $\varepsilon_{ij}(\mathbf{k},\omega)$. For each eigenmode we have expression (5.29) with $F(k)$ and $W(k)$ specified by equations (5.46) and (5.48).

5.2.1 Short-Time Interval

Let us substitute the function $F(k)$ that describes a wave packet into expression (5.29). The wavenumbers of the packet are supposed to be localized in a narrow wave number band:

$$u(x,t=0) \approx U(x)\exp(ik_0 x) \tag{5.49}$$

with $k_0 - \Delta k < k < k_0 + \Delta k$ and $\Delta k \ll k_0$. Thus $u(x,t)$ is given by equation

$$u(x,t) = \frac{1}{2\pi\Delta k}\int_{k_0-\Delta k}^{k_0+\Delta k} U(k)\exp(i(kx - W(k)t))dk, \tag{5.50}$$

where

$$U(k) = \int\limits_{-\infty}^{+\infty} U(x) \exp(-i(k - k_0)x))dx. \qquad (5.51)$$

That in turn means that the function $F(k)$ in expression (5.29) is equal to

$$F(k) = U(k)\chi(\Delta k - |k - \Delta k|). \qquad (5.52)$$

Here $\chi(x)$ is the Heaviside step-function:

$$\chi(x) = \begin{cases} 0 & \text{for} & x < 0 \\ 1 & \text{for} & x > 0 \end{cases}. \qquad (5.53)$$

Expanding the $W(k)$ function in $(k - k_0)$ we obtain

$$W(k) = W(k_0) + W'(k_0)(k - k_0) + \dots$$

$$\equiv W(k_0) + v_g(k - k_0) + \dots \equiv W(k_0) + v_g\kappa + \dots, \qquad (5.54)$$

where we have introduced $\kappa = k - k_0$.

From equation (5.50) we find

$$u(x,t) \approx \frac{U(k_0)\exp(i(k_0 x - W(k_0)t)}{2\pi\Delta k} \int\limits_{-\Delta k}^{\Delta k} \exp(i\kappa(x - v_g t))d\kappa$$

$$= \frac{U(k_0)\sin(\Delta k(x - v_g t))}{\pi\Delta k(x - v_g t)} \exp\left(ik_0\left(x - \frac{W(k_0)}{k_0}t\right)\right). \qquad (5.55)$$

This expression describes the wave packet whose phase propagates with the phase velocity $W(k_0)/k_0$, and the envelope propagates with the group velocity $v_g = W'(k_0)$, where k_0 is the initial value of the wavenumber.

5.2.2 Long Time Asymptotics

We shall consider the limit $t \to \infty$ assuming that the ratio x/t remains to be finite. To evaluate integral (5.29) we use the *stationary phase method*. The extremum of the phase, $\theta(x,t) = kx - W(k)t$, in the integrand corresponds to

$$\frac{d}{dk}(kx - W(k)t)\bigg|_{k=k_0} = 0. \qquad (5.56)$$

This leads to

$$x - W'(k_0)t = 0. \qquad (5.57)$$

It is the equation for k_0. Here k_0 is not the initial wavenumber any more , it is the extremal value of the wavenumber. The equation can be written in the form

$$v_g = W'(k_0) = \frac{x}{t}. \qquad (5.58)$$

Performing expansion of $W(k)$ around $k = k_0$ we use the relationship $x = v_g t + \delta x$, $\delta x \neq 0$ and $|\delta x| \ll 1$. For the phase $\theta(x, t)$ we find

$$\theta(x, t) = kx - W(k)t$$

$$= \kappa x - k_0 x - t\left(W(k_0) + W'(k_0)\kappa + \frac{W''(k_0)}{2}\kappa^2 + \dots\right)$$

$$= k_0 x - W(k_0)t + \kappa\delta x - \frac{W''(k_0)}{2}\kappa^2 t + \dots. \qquad (5.59)$$

Here $\kappa\delta x = tW'(k_0)(k - k_0)$ and $\kappa = (k - k_0)$. As a result we have for the integral value in expression (5.29)

$$u(x, t) \approx \frac{F(k_0)\exp(i(k_0 x - W(k_0)t))}{2\pi} \int_{-\infty}^{+\infty} \exp\left(i\left(\kappa\delta x - \frac{W''t}{2}\kappa^2\right)\right) d\kappa$$

$$= \frac{F(k_0)}{(2\pi W''(k_0)t)^{1/2}} \exp\left(i\left(k_0 x - W(k_0)t + \frac{(x - v_g t)^2}{2W''(k_0)t}\right) + i\frac{\pi}{4}\text{sign}\,W''(k_0)\right). \qquad (5.60)$$

Here we have supposed that $W''(k_0) \neq 0$. We can write equation (5.60) as

$$u(x, t) = a \exp i\theta(x, t), \qquad (5.61)$$

where a and $\theta(x, t)$ are the amplitude and eikonal respectively. According to equations (5.60) and (5.61) the wave energy density, E, is proportional to

$$E \sim a^2 \sim 1/t. \qquad (5.62)$$

We demonstrate that the wave energy density, E, obeys equation

$$\partial_t E + \partial_x(v_g E) = 0. \qquad (5.63)$$

Let us calculate the value of the wave energy contained between two points $x_1(t)$ and $x_2(t)$. It is

$$w(t) = \int_{x_1(t)}^{x_2(t)} \mathcal{E}(x, t)dx = \int_{x_1(t)}^{x_2(t)} u(x, t)u^*(x, t)dx = \int_{x_1(t)}^{x_2(t)} \frac{F(k)F^*(k)}{2\pi W(k)t}dx. \qquad (5.64)$$

If we change the x-coordinate to the k-coordinate in accordance with relationship $x = W'(k)t$ then we obtain

$$dx = tW''(k)dk. \qquad (5.65)$$

We choose the wavenumbers k_1 and k_2 to be solutions to equations

$$W'(k_1) = \frac{x_1(t)}{t} \quad \text{for} \quad k_1, \qquad (5.66)$$

and

$$W'(k_2) = \frac{x_2(t)}{t} \qquad \text{for} \qquad k_2, \tag{5.67}$$

respectively. As a result from equation (5.64) we find that the wave energy contained between two points k_1 and k_2 in the k-space

$$w(t) = \int_{k_1}^{k_2} F(k)F^*(k)dk \tag{5.68}$$

is constant for constant k_1 and k_2. That means that the wave energy contained between those two points, $x_1(t)$ and $x_2(t)$, which propagate with the group velocity, remains constant. This is a solution to the problem, since in the limit $x_2 = x_1 + dx$, $w(x,t) \approx \mathcal{E}$ and equations (5.64), (5.68) are equivalent to equation (5.63). The main contribution to the wave energy at point x and time t for $x/t =$const when $t \to \infty$ comes from the part of the initial pulse energy with $k = k_0$. It decreases as $1/t$.

5.2.3 Propagation of Electromagnetic Wave Packet in Plasmas

Let us assume that at $t = 0$ the wave packet is described by the delta-function: $u_0(x) = \delta(x)$. Then

$$u(x,t) = \frac{1}{2\pi} \int_{-\infty}^{+\infty} \cos(kx - W(k)t)dk. \tag{5.69}$$

According to formula (5.13) for the dependence of the electromagnetic wave frequency on the wavenumber, $\omega = W(k)$, we write the expression: $W(k) = (k^2c^2 + \omega_{pe}^2)^{1/2}$. Substituting this expression into (5.69) we obtain

$$u(x,t) = \frac{1}{\pi} \int_{-\infty}^{+\infty} \cos(kx)\cos\left(ct\left(k^2 + \frac{\omega_{pe}^2}{c^2}\right)^{1/2}\right)dk =$$

$$-\chi(ct - x)\frac{\omega_{pe}t}{(c^2t^2 - x^2)^{1/2}}J_1\left(\frac{\omega_{pe}}{c}(c^2t^2 - x^2)^{1/2}\right). \tag{5.70}$$

Here $J_1(x)$ is the Bessel function of the first order. This expression is nothing more that the fundamental solution of equation (5.12) which is called the Klein-Gordon equation.

The Bessel function of the first kind   The Bessel function is given by series

$$J_\nu(x) = x2^\nu \sum_{k=0}^{\infty}(-1)^k \frac{x^{2k}}{2^{2k}k!\Gamma(\nu + k + 1)}. \tag{5.71}$$

Asymptotic expansion for large value of x is

$$J_{\pm\nu}(x) = \left(\frac{2}{\pi x}\right)^{1/2} \left(\cos\left(x \pm \frac{\pi}{2}\nu - \frac{\pi}{4}\right)\right.$$

$$\left. - \sin\left(x \pm \frac{\pi}{2}\nu - \frac{\pi}{4}\right)\frac{\Gamma(\nu + 3/2)}{2x\Gamma(\nu + 1/2)}\right) + \dots. \tag{5.72}$$

The asymptotic expansion of the Bessel function for $ct \gg x$ and $t\omega_p \gg 1$ leads to

$$J_1\left(\frac{\omega_p}{c}\left(c^2t^2 - x^2\right)^{1/2}\right)$$

$$\approx \left(\frac{2c}{\pi\omega_p}\right)^{1/2} \frac{1}{\left(c^2t^2 - x^2\right)^{1/4}} \cos\left(\frac{\omega_p}{c}\left(c^2t^2 - x^2\right)^{1/2} - \frac{3\pi}{4}\right). \tag{5.73}$$

That is, for $t \to \infty$ we have

$$u(x,t) \approx \frac{(\omega_p c)^{1/2} t}{(2\pi)^{1/2}\left(c^2t^2 - x^2\right)^{3/4}} \cos\left(\frac{\omega_p}{c}\left(c^2t^2 - x^2\right)^{1/2} - \frac{3\pi}{4}\right). \tag{5.74}$$

The approach based on the stationary phase method gives the equation for k_0

$$\frac{x}{t} = W'(k_0) = \frac{k_0}{\omega_0 c^2}, \tag{5.75}$$

with a solution for k_0

$$k_0 = \frac{\omega_p}{c}\frac{x}{\left(c^2t^2 - x^2\right)^{1/2}}, \tag{5.76}$$

and for ω_0

$$\omega_0 = \frac{\omega_p}{c}\frac{c^2 t}{\left(c^2t^2 - x^2\right)^{1/2}}. \tag{5.77}$$

The second derivative of $W(k)$ with respect to k is

$$W'' = \frac{c^2}{\omega_0}\left(1 - \frac{k_0^2 c^2}{\omega_0^2}\right) = \frac{1}{\omega_p c}\frac{\left(c^2t^2 - x^2\right)^{3/2}}{t^3}. \tag{5.78}$$

Substituting these expressions into equation (5.60), we find that the obtained formula is the same as that given by equation (5.74).

The fundamental solution to the 3D Klein-Gordon equation:

$$(\partial_{tt} - c^2\Delta + \omega_0^2)u = \delta(\mathbf{x})\delta(t). \tag{5.79}$$

The fundamental solution is

$$D^r(\mathbf{x},t) = \frac{\chi(t)}{2\pi}\delta(c^2t^2 - |\mathbf{x}|^2) -$$

$$\frac{\omega_0}{4\pi}\chi(ct - |\mathbf{x}|)\frac{J_1\left(\omega_0\left(c^2t^2 - |\mathbf{x}|^2\right)^{1/2}\right)}{\left(c^2t^2 - |\mathbf{x}|^2\right)^{1/2}}. \tag{5.80}$$

5.3   Geometric Optics

Above we have represented a wave in the complex form with the complex amplitude $a(x,t)$ and the phase $\theta(x,t)$. The phase is

$$\theta(x,t) = kx - W(k)t. \tag{5.81}$$

Calculating the partial derivatives of $\theta(x,t)$ with respect to x and t we find

$$\partial_x \theta = (x - W'(k)t)\partial_x k + k, \tag{5.82}$$

$$\partial_t \theta = (x - W'(k)t)\partial_t k - W(k). \tag{5.83}$$

For the extremal value of the wavenumber, $k = k_0$, from equations (5.82) and (5.83) we obtain

$$\partial_x \theta = k, \tag{5.84}$$

$$\partial_t \theta = -W(k). \tag{5.85}$$

The phase $\theta(x,t)$ is also called the eikonal.
By virtue of cross-differentiation we have

$$\partial_{xt}\theta = \partial_{tx}\theta. \tag{5.86}$$

It gives a relationship

$$\partial_t k + \partial_x \omega = 0. \tag{5.87}$$

Since the frequency is equal to $\omega = W(k)$ equation (5.87) can be written as

$$\partial_t k + \partial_k W \partial_x k = 0, \tag{5.88}$$

or

$$\partial_t k + v_{\mathrm{g}}(k)\partial_x k = 0. \tag{5.89}$$

Below we shall discuss how to solve this type of nonlinear equations. However, by direct inspection we can see that its solution is

$$k(x,t) = k_{\mathrm{in}}(x - v_{\mathrm{g}}(k)t). \tag{5.90}$$

Here $k_{\mathrm{in}}(x)$ is the initial distribution of wavenumber k at $t = 0$. Equation (5.90) gives implicit dependence of k on t and x. It describes propagation of the wavenumber value along the characteristics $dx/dt = v_g(k)$.

   Exercise. Geometric optics implies conservation of the adiabatic invariant density, which is the action density $J = a^2/W(k)$. What is its meaning? Demonstrate that this value obeys equation

$$\partial_t J + \partial_x(v_g J) = 0. \tag{5.91}$$

   $\diamond$

   In a weakly nonuniform and slowly varying media one can write that $\omega = W(k,x,t)$. In this case from equation (5.87) it follows that

$$k = k(0) - \partial_x W\, t. \tag{5.92}$$

This describes the phase mixing, $k \to \infty$ when $t \to \infty$.

3D case.

In the spatially inhomogeneous case the generalization of equation (5.29) reads

$$u(\mathbf{x}, t) = \int \int \int F(\mathbf{k}) \exp(i(\mathbf{k} \cdot \mathbf{x} - W(\mathbf{k})t)) \frac{d^3 k}{(2\pi)^3}. \qquad (5.93)$$

Here the dispersion equation $\omega = W(\mathbf{k})$ has been used. In the limit $t \to \infty$ for $|\mathbf{x}|/t$ supposed to be finite the method of stationary phase gives equation

$$\mathbf{x} = \nabla_{\mathbf{k}} W t \equiv \mathbf{v}_g t \qquad (5.94)$$

to find $\mathbf{k}_0$.

Assuming that $|\delta \mathbf{k}| \ll |\mathbf{k}|$, where $\delta \mathbf{k} = \mathbf{k} - \mathbf{k}_0$ and $\delta \mathbf{x} = \mathbf{x} - \mathbf{v}_g t$, we find for the phase

$$\theta(\mathbf{x}, t) = \mathbf{k} \cdot \mathbf{x} - W(\mathbf{k})t = \mathbf{k}_0 \cdot \mathbf{x} + \delta \mathbf{k} \cdot \delta \mathbf{x} - W(\mathbf{k}_0)t + \frac{\hat{\mathbf{A}} \delta \mathbf{k} \cdot \delta \mathbf{k}}{2} t + \dots. \qquad (5.95)$$

Here the matrix $\hat{\mathbf{A}} = A_{ij}$ equals

$$A_{ij} = \left. \frac{\partial^2 W}{\partial k_i \partial k_j} \right|_{\mathbf{k}_0}. \qquad (5.96)$$

Further we suppose that $\det A_{ij} \neq 0$.

Substituting expression (5.95) for the phase into equation (5.93) we obtain

$$u(\mathbf{x}, t) \approx \sum_{\mathbf{k}_0} \frac{F(\mathbf{k}_0) \exp(i(\mathbf{k}_0 \cdot \mathbf{x} - W(\mathbf{k}_0)t))}{(2\pi)^3}$$

$$\times \int \exp\left(i \left(\delta \mathbf{k} \cdot \delta \mathbf{x} - \frac{\hat{\mathbf{A}} \delta \mathbf{k} \cdot \delta \mathbf{k}}{2} t \right) \right) d^3 \delta k. \qquad (5.97)$$

To calculate integral (5.97) we transform the matrix A_{ij} to principal axis with the transformation given by the matrix $B_{ij} : \delta k_i = B_{ij} \delta q_j$. We obtain

$$\left(\hat{\mathbf{A}} \delta \mathbf{k} \cdot \delta \mathbf{k} \right) = \left(\hat{\mathbf{A}} \hat{\mathbf{B}} \delta \mathbf{q} \cdot \hat{\mathbf{B}} \delta \mathbf{q} \right) = \left(\hat{\mathbf{B}}^T \hat{\mathbf{A}} \hat{\mathbf{B}} \delta \mathbf{q} \cdot \delta \mathbf{q} \right) = \sum_{j=1}^{3} \sigma_j \delta q_j^2, \qquad (5.98)$$

with $\sigma_j = \pm 1$, and reciprocal matrix equal to

$$\hat{\mathbf{A}}^{-1} = \hat{\mathbf{B}} \hat{\mathbf{B}}^T, \quad \text{and} \quad \det \hat{\mathbf{A}} |\det \hat{\mathbf{B}}|^2 = 1. \qquad (5.99)$$

Then we have

$$\int \exp\left(i \left(\delta \mathbf{k} \cdot \delta \mathbf{x} - \frac{1}{2} \left(\hat{\mathbf{A}} \delta \mathbf{k} \cdot \delta \mathbf{k} \right) t \right) \right) d^3 \delta k$$

$$= |\det \hat{\mathbf{A}}| \int \exp\left(i\left(\hat{\mathbf{B}}\delta\mathbf{q}\cdot\delta\mathbf{x} - \frac{1}{2}\left(\hat{\mathbf{A}}\cdot\hat{\mathbf{B}}\delta\mathbf{q}\cdot\hat{\mathbf{B}}\delta\mathbf{q}\right)t\right)\right)d^3\delta q$$

$$= \frac{1}{\left(\det\hat{\mathbf{A}}\right)^{1/2}}\int \exp\left(i\left(\hat{\mathbf{B}}^T\delta\mathbf{x}\cdot\delta\mathbf{q} - \frac{1}{2}\sum_{j=1}^{3}\sigma_j\delta q_j^2 t\right)\right)d^3\delta q$$

$$= \frac{1}{\left(\det\hat{\mathbf{A}}\right)^{1/2}}\prod_{j=1}^{3}\int \exp\left(i\left(B_{ji}\delta x_i\cdot\delta q_j - \frac{1}{2}\sum_{j=1}^{3}\sigma_j\delta q_j^2 t\right)\right)d\delta q_j$$

$$= \frac{\pi^{3/2}}{(2t)^{3/2}\left(\det\hat{\mathbf{A}}\right)^{1/2}}\exp\left(-i\left(\frac{\left|\hat{\mathbf{B}}^T\delta\mathbf{x}\right|^2}{2t} - \frac{\pi}{4}\sum_{j=1}^{3}\sigma_j\right)\right)$$

$$= \frac{\pi^{3/2}}{(2t)^{3/2}\left(\det\hat{\mathbf{A}}\right)^{1/2}}\exp\left(-i\left(\frac{\hat{\mathbf{A}}^{-1}\delta\mathbf{x}\cdot\delta\mathbf{x}}{2t} - \frac{\pi}{4}\sum_{j=1}^{3}\sigma_j\right)\right). \tag{5.100}$$

Here summation over repeated indices is not assumed.

Finally we have

$$u(\mathbf{x},t) \approx \sum_{\mathbf{k}_0} \frac{F(\mathbf{k}_0)\exp(i(\mathbf{k}_0\cdot\mathbf{x} - W(\mathbf{k}_0)t))}{(2\pi t)^{3/2}\left(\det\hat{\mathbf{A}}\right)^{1/2}}$$

$$\times \exp\left(-i\left(\frac{\hat{\mathbf{A}}^{-1}(\mathbf{x} - \mathbf{v}_g t)\cdot(\mathbf{x} - \mathbf{v}_g t)}{2t} - \frac{\pi}{4}\sum_{j=1}^{3}\sigma_j\right)\right). \tag{5.101}$$

5.3.1   Geometric Optics in 3D case

We represent the wave as

$$u(\mathbf{x},t) = a(\mathbf{x},t)\exp\theta(\mathbf{x},t), \tag{5.102}$$

where the eikonal is $\theta(\mathbf{x},t) = \mathbf{k}\cdot\mathbf{x} - W(\mathbf{k})t$. Hence the wave vector $\mathbf{k}$ and frequency ω equal

$$\mathbf{k} = \nabla\theta, \qquad \omega = -\partial_t\theta = -W(\mathbf{k}) \tag{5.103}$$

with curl $\mathbf{k} = 0$ and by virtue of cross-differentiation

$$\partial_t\mathbf{k} + \nabla\omega = 0. \tag{5.104}$$

The velocity of the motion of the constant phase surface with $\theta = $ const, is determined by the condition $d\theta = 0$. This gives

$$d\theta = \partial_t\theta dt + \nabla\theta d\mathbf{x} = -\omega dt + \mathbf{k}\cdot d\mathbf{x} = 0. \tag{5.105}$$

Since $\omega = (\mathbf{k} \cdot \mathbf{v}_{ph})$ we find

$$\mathbf{v}_{ph} = \frac{\mathbf{k}\omega}{|\mathbf{k}|^2} = \frac{\mathbf{k}W(\mathbf{k})}{|\mathbf{k}|^2} \tag{5.106}$$

or

$$\left(\frac{\mathbf{k}}{\omega} \cdot \frac{d\mathbf{x}}{dt}\right) = 1. \tag{5.107}$$

The value $\mathbf{k}/\omega \equiv \mathbf{k}/W(\mathbf{k})$ is a vector, it can be written as $\mathbf{k}/W(\mathbf{k}) = (1/v_{ph})$.

In a weakly inhomogeneous and nonstationary media we can write $W(\mathbf{k}, \mathbf{x}, t)$ and calculate

$$\frac{\partial k_i}{\partial t} + \frac{\partial \omega}{\partial x_i} = \partial_t k_i + \frac{\partial W}{\partial k_j}\frac{\partial k_j}{\partial x_i} + \frac{\partial W}{\partial x_i} = 0. \tag{5.108}$$

That is

$$\frac{\partial k_i}{\partial t} + v_{gj}\frac{\partial k_j}{\partial x_i} + \frac{\partial W}{\partial x_i} = 0, \tag{5.109}$$

and

$$\frac{\partial \omega}{\partial t} + v_{gj}\frac{\partial \omega}{\partial x_i} - \frac{\partial W}{\partial t} = 0. \tag{5.110}$$

The characteristics of these equations are the rays, along which

$$\mathbf{x} - \int \mathbf{v}_g dt = \text{const.} \tag{5.111}$$

Using the notation for the derivatives

$$\frac{d}{dt} = \frac{\partial}{\partial t} + v_{gj}\frac{\partial}{\partial x_i} \tag{5.112}$$

we rewrite the above obtained expressions in the form

$$\dot{x}_i = \frac{\partial W}{\partial k_i}, \qquad \dot{k}_i = -\frac{\partial W}{\partial x_i}, \qquad \dot{\omega} = \frac{\partial W}{\partial t}. \tag{5.113}$$

These are well known Hamilton equations with the canonical coordinates x and k as q and p, respectively, and W being the Hamiltonian.

Exercise. In the fluid moving with a velocity $\mathbf{U}(\mathbf{x}, t)$ the geometric optic equations have the form

$$\dot{x}_i = \frac{\partial W}{\partial k_i} + U_i, \tag{5.114}$$

$$\dot{k}_i = -\frac{\partial W}{\partial x_i} - k_j\frac{\partial U_i}{\partial x_j}, \tag{5.115}$$

$$\dot{\omega} = \frac{\partial W}{\partial t} - k_j\frac{\partial U_j}{\partial t}. \tag{5.116}$$

Use relationship

$$\omega = (\mathbf{k} \cdot (\mathbf{v}_{ph} + \mathbf{U}(\mathbf{x}, t))) \equiv W(\mathbf{k}, \mathbf{x}, t) + \mathbf{k} \cdot \mathbf{U}(\mathbf{x}, t) \tag{5.117}$$

and transform the derivative as

$$\frac{d}{dt} = \frac{\partial}{\partial t} + (v_{gj} + U_j)\frac{\partial}{\partial x_i}. \tag{5.118}$$

Find the equation that describes the energy density of waves in a stream.

◇

5.3.2 Wave Propagation in Nonhomogeneous Media

We consider the wave propagation in a stratified medium in which the speed of the wave propagation c depends only on the vertical coordinate. We assume $c = c(y)$. In the geometric optics approximation using the dispersion equation $\omega = kc$ from equations (5.113) we obtain

$$\dot{x} = \frac{k_x}{k}c, \qquad \dot{k}_x = 0,$$
$$\dot{y} = \frac{k_y}{k}c, \quad \dot{k}_y = -k\partial_y c, \tag{5.119}$$

where $k = (k_x^2 + k_y^2)^{1/2}$. Since the Hamiltonian ω does not depend on time, $kc = h$ with h being constant. Then we have $k_x = const$. Writing $k_x = k\cos\theta$ and $k_y = k\sin\theta$, we see that the angle θ of the wave ray to the horizontal is given by

$$\cos\theta = \frac{k_x}{h}c(y). \tag{5.120}$$

It is equivalent to $\cos\theta/c(y) = const$, which is nothing more than Snell's law in optics. For the $y-$component of the wavevector from $(k_x^2 + k_y^2)^{1/2}c = h$ we have

$$k_y = \left(\left(\frac{h}{c(y)}\right)^2 - k_x^2\right)^{1/2} \tag{5.121}$$

and the ray equation takes the form

$$\frac{dx}{dy} = \cot\theta = \frac{c(y)k_x}{(h^2 - k_x^2 c^2(y))^{1/2}}. \tag{5.122}$$

Therefore the ray is given by

$$x - x_0 = \int_{y_0}^{y} \frac{k_x c(y)dy}{(h^2 - k_x^2 c^2(y))^{1/2}}. \tag{5.123}$$

We may rewrite this as

$$\dot{y}^2 = c^2(y) - \left(\frac{k_x}{h}\right)^2 c^4(y), \tag{5.124}$$

which is formally the energy of the particle motion in the effective potential $c^2(y) - (k_x/h)^2 c^4(y)$.

Suppose that variations in $c(y)$ are confined to a layer $|y| < L$ with $c(y)$ being constant c_m outside the layer, $c(|y| < L) < c_m$, and with a minimum $c(0)$ at $y = 0$. From equation (5.124) we see that as $c(y)$ increases along the ray, $\cos\theta = k_x c(y)/h$ increases, θ decreases and the ray bends toward the horizontal. If $h < k_x c_m$ the ray does not leave the waveguide oscillating about the x-axis.

5.3.3 Nonadiabatic Regimes of Wave Propagation

Above we have paid attention to the fact that the geometric optics approximation is equivalent to conservation of adiabatic invariants of the wave field. One may also say that the foton (phonone, plasmon etc.) number is conserved during the wave propagation. From analysis of accuracy of the adiabatic invariant conservation we know that in nonstationary medium it implies $\dot\omega/\omega^2 \ll 1$, where ω is the oscillation frequency. The condition of the adiabatic change of the wave parameters in an inhomogeneous medium imposes a constraint on the wavenumber: $k'/k^2 \ll 1$. In the vicinity of sharp spatial discontinuity, where $k'/k^2 \gg 1$, matching of the solutions gives well known coefficients of reflection and transition of the wave. For the wave the kinematic properties of reflected and transmitted waves are given by Snell's law:

$$\frac{\sin\theta_i}{\sin\theta_r} = \frac{n_1}{n_2}, \tag{5.125}$$

where θ_i and θ_r are the angles of incidence and refraction, while n_1 and n_2 are corresponding indices of refraction. The wave is supposed to propagate from the hemiplane with the index n_1 to the hemiplane with the index n_2. During propagation in an inhomogeneous medium the wave frequency remains unchanged meanwhile the wave-vector changes: in the case of reflection and refraction of the wave at a plane interface between two media with different refraction indices, the wave vectors $\mathbf{k}_0$, $\mathbf{k}_1$ and $\mathbf{k}_2$, the wave vector of incident, reflected and refracted waves, are all different.

Now we discuss what happens with plane electromagnetic wave propagating in the medium in which the refraction coefficient changes instantaneously. In this case we have $\dot\omega/\omega^2 \gg 1$. We assume that the medium is a plasma in which a dispersion equation is given by (5.13) with ω_{pe} dependent on time. We assume that for $t < 0$ we have $\omega_0^2 = k^2 c^2$ meanwhile for $t > 0$ $\omega^2 = k^2 c^2 + \omega_{pe}^2$. At $t \leq 0$ the electric and magnetic field correspond to the plane wave, propagating in the positive direction along the x-axis:

$$\begin{aligned}
\mathbf{E} &= E_0 \cos(kx - \omega_0 t)\mathbf{e}_y = E_0 \cos(k(x - ct))\mathbf{e}_y, \\
\mathbf{B} &= B_0 \cos(kx - \omega_0 t)\mathbf{e}_z = B_0 \cos(k(x - ct))\mathbf{e}_z
\end{aligned} \tag{5.126}$$

with $E_0 = B_0$. At $t = 0$ we have $\mathbf{E} = E_0 \cos(kx)\mathbf{e}_y$ and $\mathbf{B} = B_0 \cos(kx)\mathbf{e}_z$, which provides the initial conditions for the Maxwell equations together with the initial condition for the electron velocity $\mathbf{v}(0) = 0$. As the result of the

Langmuir frequency change, for $t > 0$ we obtain

$$\mathbf{E} = (E_1 \cos(kx - \omega t) + E_2 \cos(kx + \omega t)) \, \mathbf{e}_y,$$

$$\mathbf{B} = \frac{k^2 c^2}{k^2 c^2 + \omega_{pe}^2} \left(E_1 \cos(kx - \omega t) - E_2 \cos(kx + \omega t) \right) \mathbf{e}_z + B_{st} \cos(kx) \mathbf{e}_z.$$

$$(5.127)$$

Here $\omega = (k^2 c^2 + \omega_{pe}^2)^{1/2}$ is larger than ω_0. We can also write that

$$\omega = (\omega_0^2 + \omega_{pe}^2)^{1/2}. \tag{5.128}$$

For the electron velocity in the plasma we have

$$\mathbf{v} = -\frac{ie}{m(\omega_0^2 + \omega_{pe}^2)^{1/2}} \left(E_1 \cos(kx - \omega t) - E_2 \cos(kx + \omega t) \right) \mathbf{e}_y + v_{st} \cos(kx) \mathbf{e}_y.$$

$$(5.129)$$

We see that in plasma two wave trains appear which propagate in the positive and negative directions having the amplitudes

$$E_{1,2} = \frac{E_0}{2} \left(1 \pm \frac{\omega_0}{(\omega_0^2 + \omega_{pe}^2)^{1/2}} \right). \tag{5.130}$$

In addition, the quasistatic magnetic field is generated in plasma which is equal to $\mathbf{B}_{st} = B_{st} \cos(kx) \mathbf{e}_z = -(i4\pi n e v_{st}/kc) \cos(kx) \mathbf{e}_z$. The quasistatic magnetic field amplitude is $B_{st} = (\omega_{pe}^2/(\omega_0^2 + \omega_{pe}^2)) E_0$. If the initial value of the wave frequency ω_0 is much less than final, the quasistatic magnetic field has the amplitude of the order of the initial amplitude of the field in the wave.

We can also say that the non-conservation of adiabatic invariants (the quantum number) in the case of fast change of the refraction index corresponds to creation of two waves with frequencies ω_1 and ω_2 ($\omega_{1,2} = \pm(\omega_0^2 + \omega_{pe}^2)^{1/2}$) from the wave with the frequency ω_0. The wave vector $\mathbf{k}$ does not change.

5.3.4 Wave Behavior near the Leading Edge of the Wave Train

We have assumed above that $W''(k_0) \neq 0$. Now we consider the case when there is a wavenumber, k_0, where $W''(k_0) = 0$. That means that the wave group velocity has the maximum at $k = k_0$.

The extremum of the phase $\theta = kx - W(k)t$ in the integral

$$u(x,t) = \int_{-\infty}^{+\infty} F(k) \exp(i(kx - W(k)t) \frac{dk}{2\pi} \tag{5.131}$$

is at the $k = k_0$. For k_0 we have equation

$$\frac{d\theta}{dk} = \frac{d}{dk}(kx - W(k)t) \bigg|_{k_0} = x - W'(k_0)t = 0. \tag{5.132}$$

The expansion of the phase around $k = k_0$ gives

$$\theta = k_0 x + \kappa \delta x - W(k_0)t - \frac{W'''(k_0)}{6}\kappa^3 t + \ldots, \tag{5.133}$$

where $\kappa = k - k_0$. As it follows from equations (5.131) and (5.133) for $u(x,t)$ in the limit $t \to \infty$ we have

$$u(x,t) \approx \frac{F(k_0)\exp(i(k_0 x - W(k_0)t))}{2\pi} \int\limits_{-\infty}^{+\infty} \exp\left(i\left(\kappa\delta x - \frac{W'''t}{6}\kappa^3\right)\right) d\kappa =$$

$$= \frac{F(k_0)\exp(i(k_0 x - W(k_0)t))}{\pi(4W'''(k_0)t)^{1/3}} \mathrm{Ai}\left(\frac{x - v_g t}{(W'''(k_0)t/2)^{1/3}}\right). \tag{5.134}$$

Here the Airy function, Ai(x), has asymptotics:

$$\mathrm{Ai}(x) = \begin{cases} \frac{1}{2}x^{-1/4}\exp\left(-\frac{2}{3}x^{3/2}\right) + \ldots & for \quad x \to +\infty, \\ |x|^{-1/4}\cos\left(\frac{2}{3}x^{3/2} - \frac{\pi}{4}\right) + \ldots & for \quad x \to -\infty. \end{cases} \tag{5.135}$$

When $t \to 0$ we have

$$\pi^{-1/2}t^{-1/3}\mathrm{Ai}\left(\frac{x}{t^{1/3}}\right) \to \delta(x). \tag{5.136}$$

In the vicinity of the leading edge of the train the wave has the form shown in Fig. 5.2.

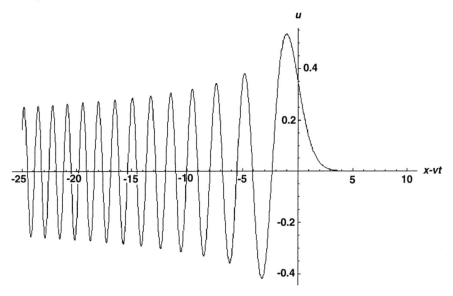

Fig. 5.2. Structure of the wave train near the leading edge.

We see that the amplitude decreases with time in this region as $t^{-1/3}$ instead $t^{-1/2}$ dependence in the region far behind the leading edge.

Exercise. Find $u(\mathbf{x}, t)$ near the leading edge in the 3D case.

◇

5.3.5 Asymptotic Expansions and the Differential Equations

The asymptotic expansions (5.60) and (5.134) obtained above are the itexact solutions to linear differential equations. To obtain formula (5.60) we have used expansion of the phase that is equivalent to the representation of the dispersion equation

$$w = w_0 + W'\kappa + \frac{1}{2}W''\kappa^2 + \dots. \tag{5.137}$$

If we neglect the remaining part of the expansion and use relationships $w \leftrightarrow i\partial_t$ and $k \leftrightarrow -i\partial_x$, the dispersion equation (5.137) leads to a differential equation for $u\exp(-iw_0 t)$:

$$i\left(\partial_t u - v_g \partial_x u\right) + \alpha \partial_{xx} u = 0. \tag{5.138}$$

This is a Schrödinger equation with $v_g = W'$ and $\alpha = W''/2$.

The dispersion equation

$$w = w_0 + W'\kappa + \frac{1}{6}W'''\kappa^3 + \dots. \tag{5.139}$$

corresponds to formula (5.134) and leads to differential equation

$$\partial_t u - v_g \partial_x u = \beta \partial_{xxx} u \tag{5.140}$$

with $v_g = W'$ and $\beta = W'''/6$. This is the linearized KdV equation.

Example. *Ion-sound Waves.* We suppose the ions in a plasma to be cold, $T_i = 0$. The equations describing the plasma motion are

$$\partial_t n + \partial_k(nv) = 0, \tag{5.141}$$

the continuity equation,

$$\partial_t v + v\partial_x v = \frac{e}{M}E, \tag{5.142}$$

the Euler equation, and

$$\partial_{xx}\varphi = 4\pi e(n_e - n) = 4\pi e\left(n_0 \exp\left(\frac{e\varphi}{T_e}\right) - n\right) \tag{5.143}$$

the Poisson equation. Here M is the ion mass, n and v are the density and velocity of the ion component, φ is the electrostatic potential. The electric field E is equal to

$$E = -\partial_x \varphi. \tag{5.144}$$

We assume that electrons have the Boltzmann distribution in the electrostatic potential φ with temperature T_e .

Linearization of equations (5.141)—(5.144) gives

$$\partial_t n^{(1)} + n_0 \partial_x v^{(1)} = 0, \tag{5.145}$$

$$\partial_t v^{(1)} = -\frac{e}{M} \partial_x \varphi^{(1)}, \tag{5.146}$$

$$\partial_{xx} \varphi^{(1)} = 4\pi e \left(\frac{n_0 e}{T_e} \varphi^{(1)} - n^{(1)} \right). \tag{5.147}$$

From these equation we obtain

$$\partial_{tt} v - \frac{T_e}{M} \partial_{xx} v = \frac{T_e}{4\pi n_0 e^2} \partial_{ttxx} v. \tag{5.148}$$

Factorization of the wave operator in the left-hand side of equation (5.148) gives

$$(\partial_t - c_{is} \partial_x)(\partial_t + c_{is} \partial_x) v = r_{De}^2 \partial_{ttxx} v \tag{5.149}$$

where $c_{is} = (T_e/M)^{1/2}$ is the ion sound velocity and $r_{De} = (T_e/ 4\pi n_0 e^2)^{1/2}$ is the electron Debye radius. Let us consider the ion sound waves that propagates in the positive direction. For such a wave $\partial_t \approx c_{is} \partial_x$. That is why equation (5.149) takes the form

$$(\partial_t + c_{is} \partial_x) v = \frac{c_{is} r_{De}^2}{2} \partial_{xxx} v. \tag{5.150}$$

This is the linear version of the KdV equation. We can incorporate the wave decay into equation (5.149) . In collisionless plasmas the wave decays are due to the Landau damping. In the k-representation the decrement of the Landau damping of the ion sound waves is equal to

$$\gamma_L = -|k| c_{is} \left(\frac{\pi}{8} \frac{m}{M} \right)^{1/2}. \tag{5.151}$$

Performing the inverse Fourier transform we obtain

$$\partial_t v + c_{is} \partial_x v - \frac{c_{is} r_{De}^2}{2} \partial_{xxx} v - c_{is} \left(\frac{\pi}{8} \frac{m}{M} \right)^{1/2} \mathcal{P} \int_{-\infty}^{+\infty} \frac{\partial_x v \, dx'}{x - x'} = 0. \tag{5.152}$$

There, in the last term of the right-hand side of the equation, is the principal value of the integral.

Chapter 6

Nonlinear Waves in Nondispersive and Dissipationless Media

Nonlinear waves in nondispersive media. Riemann waves. Euler and Lagrange variables. Wave breaking. Solution to the continuity equation.

***

The most fundamental properties of nonlinear dynamics can be seen in a system described by equations of gas dynamics in one dimensional approximation

$$\partial_t \rho + \partial_x(\rho v) = 0, \tag{6.1}$$

$$\partial_t v + v \partial_x v = -\frac{1}{\rho}\partial_x p. \tag{6.2}$$

If the gas temperature and hence the pressure vanish equations (6.1), (6.2) reduce to

$$\partial_t \rho + \partial_x(\rho v) = 0, \tag{6.3}$$

$$\partial_t v + v \partial_x v = 0. \tag{6.4}$$

This system of equations describes the gas of noninteracting particles. For example, it could be a team of sportsmen on bicycles running along a narrow (to avoid any austrip) road (see Fig. 6.1.)

Now we shall try to find a solution to the system of equations (6.3), (6.4). We assume that in zeroth order in the wave amplitude the gas is homogeneous with $\rho_0 = 1$ and with the constant in space and time velocity $v^{(0)}$.

To the first order in the wave amplitude, we have

$$\partial_t v^{(1)} + v^{(0)} \partial_x v^{(1)} = 0. \tag{6.5}$$

S. BULANOV

Fig. 6.1. Racing.

This is the simplest wave equation that corresponds to the wave with the frequency and wavenumber related to each other via the dispersion equation

$$\omega = k v^{(0)}. \tag{6.6}$$

We obtained the wave propagating in nondispersive media where both the phase velocity, ω/k, and the group velocity, $\partial\omega/\partial k$, are equal to the flow velocity $v^{(0)}$. In the frame moving with the velocity $v^{(0)}$ equation (6.5) takes the form

$$\partial_t v^{(1)} = 0. \tag{6.7}$$

Its solution is an arbitrary function of x, that does not depend on time. We choose it in the form

$$v^{(1)}(x) = a \sin kx. \tag{6.8}$$

To the second order in the wave amplitude, we obtain

$$\partial_t v^{(2)} = -v^{(1)} \partial_x v^{(1)} = a^2 k \sin kx \cos kx = -\frac{a^2 k}{2} \sin 2kx. \tag{6.9}$$

The solution to this equation,

$$v^{(2)}(x,t) = -\frac{a^2 kt}{2} \sin 2kx \tag{6.10}$$

, describes the second harmonic with the resonant growth of the amplitude in time.

To the third order in the wave amplitude, we can find that the behavior of the third harmonic is the same as the second, and so on.

In the media without dispersion high harmonics are always in resonance with the first harmonic. The resonance between harmonics appears because of the fact that the velocity of propagation is the same for all harmonics; it does not depend on the wavenumber. We see that it leads to the amplitude of the wave increasing linearly with time. The appearance and growth of high harmonics is equivalent to the steepening of the wave profile and leads to so called wave-breaking until effects of dissipation and/or dispersion do come into play.

First of all, the dissipation due to viscosity being proportional to $\nu \partial_{xx} v = -\nu k^2 v$, can lead to saturation of high harmonics. To describe these effects, we need to modify equation (6.4). Assuming that the dissipation effects have the same order as the second order in the wave amplitude perturbations, instead of equation (6.9) we find

$$\partial_t v^{(2)} - \nu \partial_{xx} v^{(2)} = -\frac{a^2 k}{2} \sin 2kx. \tag{6.11}$$

The solution to this equation reads

$$v^{(2)}(x,t) = \frac{a^2}{8\nu k}(\exp(-4\nu k^2 t) - 1)\sin 2kx. \tag{6.12}$$

For the time interval $t \ll 1/4\nu k^2$ we have $v^{(2)}(x,t) \approx -(a^2 kt/2)\sin 2kx$ in accordance with relationship (6.10). Meanwhile, when $t \to \infty$, the amplitude growth saturates as $v^{(2)}(x,t) \to (a^2/8\nu k)\sin 2kx$. If dissipation is sufficiently high to provide $a^2/8\nu k \ll a$ that is if $\nu \gg a/8k$, the second harmonic amplitude is much less than that of the first harmonic.

The dispersion effects, which are equivalent to the dependence of the phase velocity on the wave number, also can lead to saturation of the high harmonic generation. We consider the case when the dispersion effects can be described by the term $\beta \partial_{xxx} v = -i\beta k^3 v$ in the right hand side of equation (6.5). Assuming that the dispersion effects have the same order as the second order in the wave amplitude perturbations, instead of equation (6.9) we find

$$\partial_t v^{(2)} - \beta \partial_{xxx} v^{(2)} = -\frac{a^2 k}{2} \sin 2kx \tag{6.13}$$

with the solution

$$v^{(2)}(x,t) = \frac{a^2}{128\beta^2 k^2}\left((\cos(8\beta k^3 t) - 1) - 8\sin(8\beta k^3 t)\right)\sin 2kx. \tag{6.14}$$

We see that during the time interval $t \ll 1/8\beta k^3$ the amplitude of the second harmonic grows as $v^{(2)}(x,t) \approx -(a^2 kt/2)\sin 2kx$ in accordance with relationship (6.10). Meanwhile, when $t \to \infty$, the amplitude growth saturates.

Saturation of the amplitude growth in the dispersive media appears due to the propagation velocity of the second harmonic being different from $v^{(0)}$. That is the difference between the phase velocity of the first and second harmonic, and is equal to

$$\Delta v_{\text{ph}}(k) = v_{\text{ph}}(2k) - v_{\text{ph}}(k) \neq 0. \tag{6.15}$$

In this case we have the equation

$$\partial_t v^{(2)} + \Delta v_{\text{ph}}(k)\partial_x v^{(2)} = -\frac{a^2 k}{2}\sin 2kx \tag{6.16}$$

instead of the equation (6.9). Its solution

$$v^{(2)}(x,t) = \frac{a^2}{2\Delta v_{\text{ph}}(k)} \left(\cos 2kx - \cos 2k(x - \Delta v_{\text{ph}}(k)t) \right), \qquad (6.17)$$

for $t \to \infty$, does not have any singular behavior. For the time interval $t \ll 1/2k\Delta v_{\text{ph}}$ we see that $v^{(2)}(x,t) \approx -(a^2kt/2)\sin 2kx$ in accordance with relationship (6.10). Meanwhile, when $t \to \infty$, the amplitude growth saturates: the maximum of the second harmonic amplitude is equal to $v^{(2)}_{\text{max}} = a^2/\Delta v_{\text{ph}}$. It is much less than the amplitude of the first harmonic if $\Delta v_{\text{ph}} \gg a$.

The above obtained inequalities can be rewritten as $a \ll 1/2k\nu$ and $a \ll \Delta v_{\text{ph}}$, respectively. If the amplitude of the wave is small enough, the nonlinear effects are not important. Otherwise, we need to take them into account.

6.1 Simple Waves

Many years ago, Riemann obtained the exact solution to equations (6.1) and (6.2), in which the pressure was supposed to be a function of the density, $p(\rho)$, and the velocity and density were supposed to be functions of each other:

$$\rho = \rho(v), \qquad v = v(\rho). \qquad (6.18)$$

In this case, the system of the equations (6.1) — (6.2) can be rewritten as

$$(\partial_t v + v\partial_x v)\,\partial_v \rho = -\rho\partial_x v, \qquad (6.19)$$

$$\partial_t v + v\partial_x v = -\frac{c_s^2(\rho)}{\rho}\partial_v \rho \partial_x v. \qquad (6.20)$$

Here the speed of sound is equal to $c_s(\rho) = (\partial p/\partial \rho)^{1/2}$. These two equations are consistent when

$$(\partial_\rho v)^2 = \frac{c_s^2(\rho)}{\rho^2}, \qquad (6.21)$$

that is when

$$v(\rho) = \pm \int \frac{c_s(\rho)}{\rho}d\rho. \qquad (6.22)$$

This expression gives a relationship between v and ρ. Substituting (6.22) into equation (6.20) we find

$$\partial_t v + (v \pm c_s(v))\partial_x v = 0. \qquad (6.23)$$

For the process with the pressure depending on the density:

$$p(\rho) = p_0 \left(\frac{\rho}{\rho_0} \right)^\gamma, \qquad (6.24)$$

the speed of sound depends on ρ as

$$c_s(\rho) = c_0 \left(\frac{\rho}{\rho_0}\right)^{(\gamma-1)/2}. \qquad (6.25)$$

By virtue of equations (6.22) and (6.25) we have

$$c_s(v) = c_0 + \frac{\gamma - 1}{2} v. \qquad (6.26)$$

As a result we find

$$\partial_t v + \left(c_0 \pm \frac{\gamma + 1}{2} v\right) \partial_x v = 0. \qquad (6.27)$$

In the frame of coordinates moving with the speed c_0 for the function $u = \pm(\gamma + 1)v/2$, we obtain

$$\partial_t u + u \partial_x u = 0, \qquad (6.28)$$

which is formally the same as equation (6.4). We can say that it describes the flow of noninteracting phonons. To solve this equation we need to learn the Euler and Lagrange coordinates.

6.1.1 The Lagrange Coordinates

The Euler coordinates are x and t. The Lagrange coordinates can be defined in many different ways. We shall use the Lagrange coordinates, x_0 and t', where $t' = t$ and x_0 is the initial coordinate of the fluid element (at $t = 0$) (See Fig. 6.2).

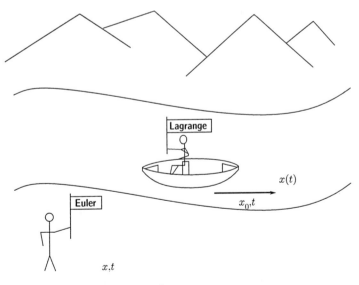

Fig. 6.2. Euler and Lagrange coordinates.

The change of the coordinates

$$t = t', \qquad x = x_0 + \xi(x_0, t'), \tag{6.29}$$

gives the relationship between the Euler and Lagrange coordinates. Here $\xi(x_0, t')$ is a displacement of the fluid element from its initial position, x_0, to the point x during time t. The fluid velocity is

$$v = \partial_t \xi. \tag{6.30}$$

According to well known rules we calculate expressions for the partial derivatives

$$(\partial_t)_{\text{Euler}} = (\partial_t)_{\text{Lagrange}} + (\partial_{x_0})_{\text{Lagrange}} \, \partial_t x_0 |_{x=\text{const}}, \tag{6.31}$$

$$(\partial_x)_{\text{Euler}} = (\partial_{x_0})_{\text{Lagrange}} \, \partial_x x_0 |_{t=\text{const}}. \tag{6.32}$$

From equations (6.29) and (6.31) we find

$$\partial_t x_0 = -\frac{v}{1 + \partial_{x_0} \xi}, \tag{6.33}$$

and the Jacobian of the transformation is

$$J = |\partial_{x_0} x| = |1 + \partial_{x_0} \xi|. \tag{6.34}$$

As a result we obtain

$$(\partial_t)_{\text{Euler}} = (\partial_t - v \partial_x x_0 \partial_{x_0})_{\text{Lagrange}}, \tag{6.35}$$

$$(\partial_t + v \partial_x)_{\text{Euler}} = (\partial_t)_{\text{Lagrange}}. \tag{6.36}$$

Using these formulas we find representation of the equation (6.4) or (6.28) in the Lagrange coordinates:

$$\partial_t v = 0, \qquad \text{or} \qquad \partial_{tt} \xi = 0. \tag{6.37}$$

It has the solutions

$$v(x_0, t) = v_0(x_0), \tag{6.38}$$

$$\xi(x_0, t) = v_0(x_0) t. \tag{6.39}$$

For the Jacobian we find

$$J = |\partial_{x_0} x| = |1 + \partial_{x_0} \xi| = |1 + v_0'(x_0) t|. \tag{6.40}$$

When $v_0'(x_0) < 0$ the Jacobian tends to zero while $t \to t_{\text{br}}$ where

$$t_{\text{br}} = \frac{1}{|v_0'(x_0)|_{\text{max}}} \tag{6.41}$$

with $|v_0'(x_0)|_{\text{max}}$ being the maximum value of the velocity gradient at $t = 0$. The singularity that is reached at $t = t_{\text{br}}$ corresponds to the wave breaking.

The wave breaking is a fundamental phenomena of generic nature. You can see the wave breaking when a wave reaches the sea shore. The wave breaking depending on particular circumstances leads either to the shock wave formation, solitary wave generation, or multistream kinetic motion of particles.

The development of the wave break is shown in Fig. 6.3.

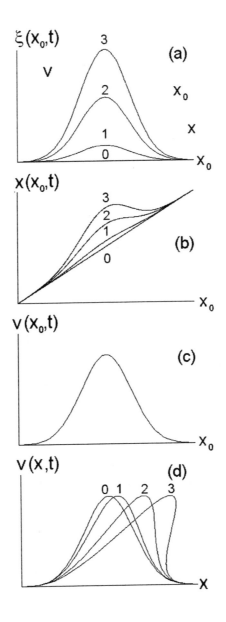

Fig. 6.3. Wave breaking.

What happens after the wave breaking? First of all, after the break equation (6.4) or (6.28) is not more valid. We need to take into account the dissipation that prevents the break and results in the shock wave formation.

Exercise. Find the solution to the kinetic equation for the distribution function $f(x, v, t)$,

$$\partial_t f + v \partial_x f = 0, \tag{6.42}$$

that describes noninteracting particles in the case when initial distribution is $f(x, v, 0) = \delta(v - v_0(x_0))$.

Characteristics of equation (6.42) obey equations $dx/dt = v$, $dv/dt = 0$,. Hence, $x = x_0 + vt$, $v = v_0$. The solution to the kinetic equation is $f = f_0(x - vt, v)$. For given initial conditions we have $f = \delta(v - v_0(x - vt))$.

◇

Exercise. Find solution to the kinetic equation

$$\partial_t f + v \partial_x f = -\nu f \tag{6.43}$$

with $\nu = $ const. Investigate the wave breaking.

◇

6.1.2   Solution to the Continuity Equation

The continuity equation is

$$\partial_t \rho + \partial_x (\rho v) = 0. \tag{6.44}$$

Exercise. Solve equation (6.44) in the Euler coordinates.

◇

◇

We shall use the Lagrange coordinates. We rewrite equation (6.44) as

$$\partial_t \rho + v \partial_x \rho == -\rho \partial_x v, \tag{6.45}$$

or

$$\partial_t \ln \rho + v \partial_x \ln \rho = -\partial_x v. \tag{6.46}$$

In the Lagrange coordinates this equation takes the form

$$\partial_t \ln \rho = -\partial_{x_0} v \partial_x x_0 = -\frac{\partial_{x_0} v}{(1 + \partial_{x_0} \xi)}$$

$$= -\frac{\partial_{t x_0} \xi}{(1 + \partial_{x_0} \xi)} = -\partial_t \ln(1 + \partial_{x_0} \xi). \tag{6.47}$$

Integrating the equation with respect to time, we obtain

$$\rho(x_0, t) = \frac{\rho(x_0, 0)}{1 + \partial_{x_0} \xi}, \tag{6.48}$$

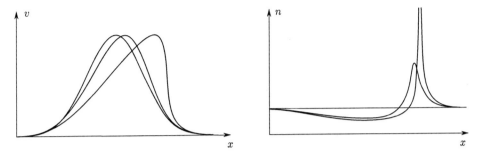

Fig. 6.4. Density behavior during wave breaking.

where $\rho(x_0, 0)$ is the initial distribution of the density. Singularity in the flow appears when the Jacobian vanishes: $J(x_0, t) = |1 + \partial_{x_0}\xi| = 0$. We see from equation (6.48) that the density there becomes infinite. (See Fig. 6.4).

Exercise. Find the behavior of the density when the Riemann wave breaks.

◇

Exercise. According to equation (6.48) the density tends to infinity when $\partial_{x_0}\xi = -1$. Find the value of the mass integrated over the wave breaking region. Is the mass finite?

◇

6.2   Nonlinear Wave Breaking

Different types of the wave breaking. Nonlinear langmuir waves in a cold plasma. Nonlinear caustics. The hodograph transformation.

***

6.2.1   Different Types of the Wave Breaking. Nonlinear Langmuir Waves in a Cold Plasma.

We consider one-dimensional problems. Let us investigate the motion of the electron component in collisionless plasmas. We assume that plasma is cold ($T_e = 0$) and ions due to the smallness of the mass ratio ($m/M \ll 1$) are supposed to be at rest with the density $n_0(x)$.

The density, $n(x, t)$, and the velocity, $v(x, t)$, obey equations

$$\partial_t n + \partial_x(nv) = 0, \tag{6.49}$$

$$\partial_t v + v\partial_x v = -\frac{e}{m}E, \tag{6.50}$$

where E is the electric field. The self-consistent electric field in the plasma must be found from the Poisson equation

$$\partial_x E = 4\pi e(n_0 - n). \tag{6.51}$$

From this equation it follows that

$$n = n_0 - \frac{1}{4\pi e}\partial_x E. \tag{6.52}$$

Substituting n into equation (6.51) we find

$$-\frac{1}{4\pi e}\partial_{xt}E + \partial_x\left(\left(n_0 - \frac{1}{4\pi e}\partial_x E\right)v\right) = 0. \tag{6.53}$$

Integrating this equation on x we obtain

$$\partial_t E + v\partial_x E = 4\pi e n_0 v. \tag{6.54}$$

The dependence of the ion density $n_0(x)$ on coordinate is supposed to be given. Here x is the Euler coordinate. In the Lagrange coordinates equations (6.50) and (6.54) can be written as

$$\partial_t v = -\frac{e}{m}E, \tag{6.55}$$

and

$$\partial_t E = 4\pi e n_0(x_0 + \xi(x_0, t))\partial_t\xi. \tag{6.56}$$

Integrating equation (6.56) on t we find

$$E = 4\pi e \int_0^{\xi(x_0,t)} n_0(x_0 + s)\, ds. \tag{6.57}$$

Substituting this expression into equation (6.55) we obtain finally

$$\partial_{tt}\xi + \frac{4\pi e^2}{m} \int_0^{\xi(x_0,t)} n_0(x_0 + s)\, ds = 0. \tag{6.58}$$

This is a nonlinear ordinary differential equation, since the variables depend on the coordinate x_0 as on the parameter.

If the density of ions does not depend on the coordinate, $n_0 = $ const, equation (6.58) becomes linear:

$$\partial_{tt}\xi + \omega_{pe}^2\xi = 0. \tag{6.59}$$

The solution to this equation,

$$\xi(x_0, t) = A(x_0)\cos\omega_{pe}t + B(x_0)\sin\omega_{pe}t, \tag{6.60}$$

where $A(x_0)$ and $B(x_0)$ are arbitrary functions of the coordinate x_0, describes the finite amplitude Langmuir oscillations. The frequency of the oscillations

$$\omega_{pe} = \left(\frac{4\pi n_0 e^2}{m}\right)^{1/2} \tag{6.61}$$

does not depend on the oscillation amplitude. The nonlinearity of the process under consideration is hidden inside the relationship between the Euler and Lagrange coordinates given by $x = x_0 + \xi(x_0, t)$.

We can choose arbitrary functions $A(x_0)$ and $B(x_0)$ to obtain the solution (6.60) in the form

$$\xi(x_0, t) = a \sin(kx_0 - \omega_{pe}t).$$ (6.62)

This expression describes the Langmuir wave that propagates to the right along the x_0-axis with a constant phase velocity $v_{ph} = \omega_{pe}/k$. Since $v = \partial_t \xi$ we have from equation (6.62)

$$v(x_0, t) = -a\omega_{pe} \cos(kx_0 - \omega_{pe}t).$$ (6.63)

The wave breaking condition is $\xi' \leq -1$, that is $ka \cos(kx_0 - \omega_{pe}t) \leq -1$. We see the wave breaking of the Langmuir wave occurs when $ka \geq 1$.

The wave break has a generic local structure that can be described as it follows. Since the Jacobian $J = |\partial x/\partial x_0|$ vanishes at the wave break we can expand it as

$$J \approx 1 + v'(x_0)t \approx 1 + v'_m \left(1 - \frac{k^2 x_0^2}{2}\right) t.$$ (6.64)

When $t = 1/v'_m$ the Jacobian vanishes at $x_0 = 0$. For $v(x_0, t)$ and $\xi(x_0, t)$ we have approximately

$$v(x_0, t) \approx v'_m (x_0 - k^2 x_0^3/6),$$ (6.65)

and

$$\xi(x_0, t) \approx v(x_0, t)t \approx vt.$$ (6.66)

Since $x = x_0 + \xi(x_0, t)$ we can write the relationship between the velocity and the Euler coordinate near the wave break

$$x \approx v\delta t - \frac{k^2}{6} \left(\frac{v}{v'_m}\right)^3.$$ (6.67)

Here $\delta t = t - 1/|v'_m|$. We see that the velocity gradient, $\partial_x v$, tends to infinity at $x = 0$ for $\delta t \to 0$.

Now we consider the case when $ka = 1$. It means that the wave is at the threshold of the break, where $\partial_{x_0} v = 0$. The threshold can be reached by slow increasing of the wave amplitude or/and by decreasing of the wavenumber. Near the threshold of the wave breaking the structure of the wave is different from that described by formula (6.67). This is due to two constrains imposed by conditions:

$$\partial_{x_0}\xi = -1, \quad \text{and} \quad \partial_{x_0} v = 0, \quad \text{or} \quad \partial_{x_0 t}\xi = 0.$$ (6.68)

For the wave phase we write

$$kx_0 - \omega_{pe}t = \pi + k\delta,$$ (6.69)

that is

$$x_0 = \frac{\pi}{k} + v_{\mathrm{ph}}t + \delta. \tag{6.70}$$

Here δ is a measure of the wavebreaking region size. It is supposed to be small compared to the wavelength value. The expressions given by (6.62) and (6.63) take the form

$$\xi(x_0, t) = -a \sin k\delta, \tag{6.71}$$

and

$$v(x_0, t) = a\omega_{\mathrm{pe}} \cos k\delta, \tag{6.72}$$

respectively. The conditions (6.68), $\xi' = -1$ and $v' = 0$, give $\delta = 0$ and $ka = 1$.

Near the wave breaking point we expand ξ and v in δ :

$$\xi(x_0, t) \approx -a \left(k\delta - \frac{(k\delta)^3}{6} + \dots \right), \tag{6.73}$$

and

$$v(x_0, t) \approx a\omega_{\mathrm{pe}} \left(1 - \frac{(k\delta)^2}{2} + \dots \right). \tag{6.74}$$

Now we can calculate the dependence of the velocity v on the Euler coordinate x in explicit form. As it follows from equations (6.70) and (6.73) the Euler coordinate x as a function of δ is

$$x = x_0 + \xi(x_0, t) = \frac{\pi}{k} + v_{\mathrm{ph}}t + \delta - ak\delta + a\frac{(k\delta)^3}{6} = \frac{\pi}{k} + v_{\mathrm{ph}}t + a\frac{(k\delta)^3}{6}. \tag{6.75}$$

Expressing δ as a function of x we introduce it into equation (6.74). As a result we have

$$v(x_0, t) \approx a\omega_{\mathrm{pe}} \left(1 - \frac{1}{2} \left(\frac{6(x - \pi/k - v_{\mathrm{ph}}t)}{a} \right)^{2/3} \right). \tag{6.76}$$

This demonstrates (x, v), a cusp-like structure of the wavebreak with generic dependence $v \sim x^{2/3}$ in the phase plane.

We have studied two types of the wave break. The first regime leads to the singularity in the phase plane, (x, v), that can be described by formula $x = tv - v^3$; the second regime is described by the function $x = v^{3/2}$. These two structures of singularities correspond to two first catastrophes, the fold and the cusp, in the catastrophe theory.

Exercise. In the inhomogeneous plasma the Langmuir frequency ω_{pe} depends on the coordinate. Since

$$\partial_t k = -\partial_x \omega_{\mathrm{pe}} \tag{6.77}$$

the wavelength changes. Demonstrate this with the help of equation (6.58) assuming the displacement ξ to be small. Find the break structure.

◇

Exercise. When the Langmuir oscillations are driven by external electric field, $E_0 \cos \omega_0 t$, we can write

$$\ddot{\xi} + \omega_{pe}^2(x_0)\xi = -\frac{e}{m}E_0 \cos \omega_0 t. \tag{6.78}$$

Investigate wave breaking in the plasma resonance region where $\omega_{pe}(x_0) = \omega_0$.

$\diamond$

6.2.2 Nonlinear Caustics.

In the 2-D and 3-D cases the wave breaking acquires novel features. We consider the plasma wave whose constant phase surfaces can be described by expression

$$\psi(x, y, z) = \omega_p(y, z) \left(t - \frac{x}{v_{ph}} \right) = const. \tag{6.79}$$

Here the plasma wave frequency is approximated in the vicinity of the axis by the parabolic form

$$\omega_p(y) \approx \omega_p(0) + \Delta\omega_p \left(\left(\frac{y}{S_y} \right)^2 + \left(\frac{z}{S_z} \right)^2 \right). \tag{6.80}$$

The transverse inhomogeneity of ω_p is caused by the inhomogeneity of the plasma density. Thus we can write the constant phase curves as

$$x_0 \equiv x - v_{ph}t + \frac{\psi v_{ph}}{\omega_p(0)} = \frac{y_0^2}{2R_y} + \frac{z_0^2}{2R_z}, \tag{6.81}$$

where R_y and R_z are the curvature radii in the $y-$ and $z-$ directions.

The real position of the constant phase curves in a nonlinear plasma wave is shifted from the curves given above by the oscillation amplitude ξ. Thus, when $|\mathbf{R}|$ becomes of the order of the electron displacement ξ, the wake plasma wave starts to break. A number of important features about the geometry of the phase surfaces near the axis at wave break can be understood by taking, for simplicity, the displacement to be perpendicular to the phase surfaces (in the 3-D case the paraboloid surfaces $x_0 - y_0^2/2R_y - z_0^2/2R_z = const$) derived in the linear approximation, and writing the new surface as

$$x = x_0 + \xi_x(x_0, y_0, z_0) = \frac{y_0^2}{2R_y} + \frac{z_0^2}{2R_z} + \frac{\xi(y_0, z_0)}{(1 + (y_0/R_y)^2 + (z_0/R_z)^2)^{1/2}}, \tag{6.82}$$

$$y = y_0 + \xi_y(x_0, y_0, z_0) = y_0 \left(1 + \frac{\xi(y_0, z_0)}{R_y (1 + (y_0/R_y)^2 + (z_0/R_z)^2)^{1/2}} \right), \tag{6.83}$$

$$z = z_0 + \xi_z(x_0, y_0, z_0) = z_0 \left(1 + \frac{\xi(y_0, z_0)}{R_z (1 + (y_0/R_y)^2 + (z_0/R_z)^2)^{1/2}} \right). \tag{6.84}$$

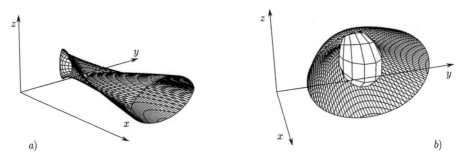

Fig. 6.5. a) A surface of constant phase in the 3-D case for $R \leq \xi$ and $\xi = const$. b) A surface of constant phase in the 3-D case for $R \leq \xi$ and $\xi \neq const$ with $\delta = 4$.

If we neglect the dependence of the displacement $\xi(y_0, z_0)$ on the coordinate y_0, z_0 along the wave front, Eqs. (6.82-6.84) define a so-called surface parallel to a paraboloid. The singularity that is formed for $\min\{R_y, R_z\} \leq \xi_m$ corresponds to the self-intersection of the electron trajectories. It is shown in Fig. 6.5 a.

In a more realistic analysis the amplitude of the displacement is not constant and in the most likely conditions it has its maximum on the axis. We describe this dependence with the Lorentzian form:

$$\xi(y_0, z_0) = \frac{\xi_m}{1 + (y_0/S_y)^2 + (z_0/S_z)^2}. \tag{6.85}$$

The singularity that is formed for $R \leq \xi_m$ is shown in Fig. 6.5 b.

To clarify the singularity structures we consider the 2-D case where $R_z \to \infty$ and $S_z \to \infty$. Then the constant phase surface has a parabolic form, which corresponds to our 2D computer simulations discussed below.

If ξ is independent of the coordinate y_0, we obtain the singularity that is shown in Fig. 6.6 a, where the form of the wave front in the x, y plane is presented for increasing amplitudes of the displacement ξ.

Near this singularity the dependence of x on y is of the form $y \approx |x|^{3/4}$. For $R < \xi_m$ a multivalued structure appears. It is known as the "swallow tail" in catastrophe theory. Near the breaking threshold, the value of the displacement on the axis is close to that of the curvature radius, $\xi(0)/R - 1 = \varepsilon \ll 1$, and the size of the swallowtail is of order ε^2 along x and $\varepsilon^{3/2}$ along y. The typical portion of the affected wave measured in terms of y_0 is $\varepsilon^{1/2}$.

In the case of the displacement with its maximum on the axis we write

$$\xi(y_0) = \frac{\xi_m}{1 + (y_0/S_y)^2}. \tag{6.86}$$

Below the threshold, when $\varepsilon < 4\delta$, where $\delta = (R_y/S_y)^2$, the type of the singularity changes as shown in Fig. 6.6 b. At the wave-breaking threshold, the singularity is described by the generic semi-cubic form $x \approx |y|^{2/3}$. This corresponds to the catastrophe that, in the space of the parameters x, y and ξ, is

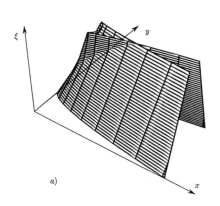
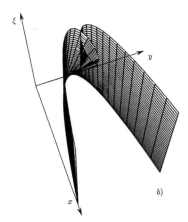

Fig. 6.6. a) The form of the constant phase surface in the 2-D case for growing amplitudes of the displacement ξ when ξ does not depend on the coordinates. b) The form of the constant phase surface in the 2-D case for growing amplitudes of the displacement ξ when ξ has a Lorentzian form.

known as the "Whitney's umbrella". For $\varepsilon = \xi(0)/R_y - 1$ well above the breaking threshold, we have $\varepsilon \approx 4\pi\xi\Delta\omega_p/(\omega_p(0)k_l S_y^2)$, while $\delta \approx (\xi/S_y)^2$. Then $\varepsilon > \delta$ and from Eqs. (6.82-6.83) it follows that the type of singularity that is realized corresponds to the "swallow tail" shown in Fig. 6.6 a.

6.2.3   The Hodograph Transformation.

There is one more method to find the exact solutions to nonlinear equations in partial derivatives. This is the hodograph transform applied to quasilinear differential equations.

We consider equations of gas dynamics that can be written in the form

$$\partial_t c_s + v\partial_x c_s + \frac{\gamma - 1}{2}c_s\partial_x v = 0, \tag{6.87}$$

$$\partial_t v + v\partial_x v + \frac{2}{\gamma - 1}c_s\partial_x c_s = 0. \tag{6.88}$$

It has been supposed that the pressure, p, depends on the density, ρ , only. The speed of sound is $c_s = (\partial p/\partial \rho|s)^{1/2}$. This is the isentropic flow. Equations (6.87), (6.88) are linear with respect to derivatives ∂_t and ∂_x, that is why they are called the quasilinear differential equations. They can be transformed into a set of linear equations by interchanging the dependent, c_s and v, and independent, x and t, variables. Instead of variables $c_s(x,t)$ and $v(x,t)$ we shall use $x(v, c_s)$ and $t(v, c_s)$. The relationships between those variables

$$\partial_v x = -\frac{1}{J}\partial_t c_s, \qquad \partial_{c_s} x = \frac{1}{J}\partial_t v, \tag{6.89}$$

$$\partial_v t = \frac{1}{J}\partial_x c_s, \qquad \partial_{c_s} t = -\frac{1}{J}\partial_x v, \tag{6.90}$$

contain the Jacobian of the transformation

$$J = \{v, c_s\} \equiv \partial_t v \partial_x c_s - \partial_t c_s \partial_x v. \tag{6.91}$$

Here

$$\{f, g\} = \partial_t f \partial_x g - \partial_t g \partial_x f \tag{6.92}$$

are the Poisson brackets.

Exercise. Demonstrate that the Riemann wave is not described in the frame of this transform.

◇

Then we substitute expressions given by relationships (6.89), (6.90) into equations (6.87), (6.88). Since these equations are quasilinear the Jacobian cancels through. As a result we obtain

$$\partial_v x - v \partial_v t + \frac{\gamma + 1}{2} c_s \partial_{c_s} t = 0, \tag{6.93}$$

$$\partial_{c_s} x - v \partial_{c_s} t + \frac{2}{\gamma - 1} c_s \partial_v t = 0. \tag{6.94}$$

These are linear equations.

Eliminating x by cross-differentiation we obtain the equation for one function, $t(v, c_s)$:

$$\partial_{c_s c_s} t + \frac{\gamma + 1}{2 c_s} \partial_{c_s} t - \frac{2}{\gamma - 1} \partial_{vv} t = 0. \tag{6.95}$$

We introduce a new variable μ and parameter κ equal to

$$\mu = \frac{2 c_s}{\gamma - 1}, \quad \text{and} \quad \kappa = \frac{1}{2} \left(\frac{\gamma + 1}{\gamma - 1} \right), \tag{6.96}$$

respectively. In variables μ and v equation (6.95) takes the form

$$\partial_{\mu\mu} t + \frac{2\kappa}{\mu} \partial_\mu t - \partial_{vv} t = 0. \tag{6.97}$$

This equation is of a hyperbolic type. For $\kappa = 1$ it has a form of the wave equation in the spherical coordinates. For κ being integer its solution can be expressed in the form

$$t = (\partial_{vv} - \partial_{\mu\mu})^{\kappa - 1} \left(\frac{f(v + \mu) + g(v - \mu)}{\mu} \right), \tag{6.98}$$

where f and g are arbitrary functions.

Chapter 7

Nonlinear Aspects of Hydrodynamic Type Instabilities

Nonlinear aspects of hydrodynamic type instabilities. Equations of motion of incompressible fluid. Theory of shallow water. The Rayleigh-Taylor instability of thin layer of water. The Rayleigh-Taylor instability of compressible shell. Tearing mode instability of thin current sheet. Self-similar solutions. Nonlinear stage of the instability of electron beam in a plasma. Some remarks on the behavior of unstable media. Jean's instability.

***

7.1 Equations of Motion of Incompressible Fluid

Here we write a set of hydrodynamics equations. The fluid is supposed to be incompressible. That corresponds to condition

$$\operatorname{div} \mathbf{v} = 0. \tag{7.1}$$

The equation of motion is the Euler equation

$$\partial_t \mathbf{v} + (\mathbf{v} \cdot \nabla)\mathbf{v} = -\frac{\nabla p}{\rho} - g\mathbf{e}_y. \tag{7.2}$$

Here g is the gravitational acceleration. Writing second term in the right hand side of the equation in the form

$$(\mathbf{v} \cdot \nabla)\mathbf{v} = \omega \times \mathbf{v} + \frac{1}{2}\nabla v^2, \tag{7.3}$$

where $\boldsymbol{\omega} = \text{curl } \mathbf{v}$, and applying the curl operation to equation (7.2) we obtain

$$\partial_t \boldsymbol{\omega} + (\mathbf{v} \cdot \nabla)\boldsymbol{\omega} - (\boldsymbol{\omega} \cdot \nabla)\mathbf{v} = 0, \tag{7.4}$$

which can be written also in the form

$$\partial_t \boldsymbol{\omega} = \nabla \times (\mathbf{v} \times \boldsymbol{\omega}). \tag{7.5}$$

This equation describes a property of the vector field $\boldsymbol{\omega}$ to be frozen into fluid motion. It means that the flux of the vector $\boldsymbol{\omega}$ through the surface S with the boundary ∂S moving with the fluid is constant, $\oint_{\partial S}(\boldsymbol{\omega} \cdot \mathbf{n})dS =$const. Here $\mathbf{n}$ is a normal to the surface S vector. Equation (7.5) has an exact solution which can be written in the Lagrange coordinates.

In the 3D case we choose the Lagrange coordinates as it follows. The elementary liquid volume moves having the Euler coordinate $\mathbf{x}(\mathbf{x}^0, t)$, with $\mathbf{x}^0$ being the initial position of the volume. At $t = 0$ we have

$$\mathbf{x}(\mathbf{x}^0, 0) = \mathbf{x}^0, \qquad \left(\frac{\partial x_i}{\partial x_j^0}\right)_{t=0} = \delta_{ij}. \tag{7.6}$$

For the differential $d\mathbf{x}$ we find

$$dx_i = M_{ij}dx_j^0, \qquad \text{with} \qquad M_{ij} = \frac{\partial x_i}{\partial x_j^0}. \tag{7.7}$$

The material curve (line), C_L, comprises points of the fluid. If p is the parameter of the curve then $\mathbf{x} = \mathbf{x}(\mathbf{x}^0(p), t)$. In the Lagrange coordinates we have $\partial M_{ij}/\partial t = \partial u_i/\partial x_j^0$, and

$$\frac{\partial dx_i}{\partial t} = dx_j^0 \frac{\partial u_i}{\partial x_j^0} = (d\mathbf{x} \cdot \nabla)\mathbf{u}. \tag{7.8}$$

The solution to equation (7.5) has been found by Cauchy. It is

$$\omega_i(\mathbf{x}_0, t) = \frac{\partial x_i}{\partial x_{0,j}} \frac{\omega_j(\mathbf{x}_0, 0)}{D(\mathbf{x}_0, t)}, \tag{7.9}$$

where $D = \det(M_{ij})$ is the Jacobian of the transformation from the Euler to Lagrange coordinates and $\omega_j(\mathbf{x}_0, 0)$ gives the initial distribution of the vector field ω_j. In incompressible fluid $D = 1$.

If $\boldsymbol{\omega}(\mathbf{x}_0, 0)$ at $t = 0$ then it remains to be equal to zero. The flow under such a condition is potential, that is $\mathbf{v} = -\nabla\varphi$, where φ is a scalar potential. From equation (7.1) it follows that

$$\Delta\varphi = 0 \tag{7.10}$$

$$\partial_t\varphi - \frac{1}{2}(\nabla\varphi)^2 = -\frac{p - p_0}{\rho} - gy. \tag{7.11}$$

7.1.1   Theory of Shallow Water

We consider the gravitational waves on the surface of the shallow water of finite depth, h. We need to impose the boundary conditions at the water surface, $y = \eta(x,t)$, and at the bottom, $y = -h$. On the water surface at $y = \eta(x,t)$ we have $p = p_0$, then the equation (7.11) takes the form

$$\partial_t \varphi - \frac{1}{2}(\nabla \varphi)^2 = -g\eta. \tag{7.12}$$

At $y = -h$ the normal component of the velocity vanishes:

$$v_y = 0, \qquad \text{that is} \qquad \partial_y \varphi = 0. \tag{7.13}$$

The kinematic condition at the surface gives

$$\frac{d\eta}{dt} = v_y \tag{7.14}$$

for $y = \eta(x,t)$, that is $\dot{\eta} = -\partial_y \varphi$. Since $\dot{\eta} = \partial_t \eta + (\mathbf{v}\cdot\nabla)\eta$ we have

$$\partial_t \eta - (\nabla \varphi \cdot \nabla \eta) = -\partial_y \varphi. \tag{7.15}$$

Linearization of equations (7.12) and (7.15) leads to

$$\partial_t \varphi = -g\eta, \qquad \text{and} \qquad \partial_t \eta = -\partial_y \varphi \tag{7.16}$$

for $y = 0$.

For $\varphi = \varphi(y)\exp(-i(\omega t - kx))$ from equation (7.10) we obtain

$$\frac{d^2\varphi}{dy^2} - k^2\varphi = 0 \tag{7.17}$$

with

$$\frac{d\varphi}{dy} = i\omega\eta \qquad \text{and} \qquad \eta = \frac{i\omega}{g}\varphi \tag{7.18}$$

for $y = 0$, and

$$\frac{d\varphi}{dy} = 0 \tag{7.19}$$

for $y = -h$.

The solution to equation (7.17) is

$$\varphi = -i\frac{g}{\omega}\frac{\cosh k(y + h)}{\cosh kh}\exp\left(-i(\omega t - kx)\right). \tag{7.20}$$

From boundary conditions at the water surface given by (7.18), written in the form $d\varphi/dy = -(\omega^2/g)\varphi$, we obtain the dispersion equation

$$\omega^2 = gk\tanh kh. \tag{7.21}$$

In the deep water, when the water depth is much larger that the wavelength of the wave: $h \gg k^{-1} \equiv \lambda/2\pi$, from (7.21) we obtain

$$\omega(k) = (gk)^{1/2}. \tag{7.22}$$

In the shallow water with $h \ll k^{-1} \equiv \lambda/2\pi$ we have

$$\omega(k) = k(gh)^{1/2} + \frac{k^3 h^2 (gh)^{1/2}}{6} + \dots, \tag{7.23}$$

with the dispersion corresponding to the linear form of the KdV equation. From equation (7.23) we have that the phase velocity,

$$v_{\text{ph}}(k) = (gh)^{1/2} \left(1 + \frac{(kh)^2}{6} + \dots \right), \tag{7.24}$$

is less where the water depth, h, is smaller. That explains why the "tsunami" reaches the islands in a long chain of islands almost at the same time.

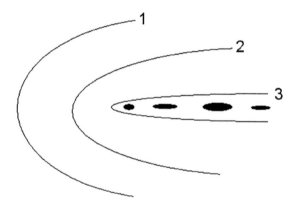

Fig. 7.1. Gravitation waves near the chain of islands.

Exercise. Investigate the gravitational waves on the surface between two fluids with densities ρ_1 and ρ_2, moving with velocities U_1 and U_2 in the x-direction. Demonstrate that the dispersion equation for two semi-infinite fluids in the y-direction is

$$\frac{\omega(k)}{k} = \frac{\rho_1 U_1 + \rho_2 U_2}{\rho_1 + \rho_2} \pm \left(\frac{g}{k} \left(\frac{\rho_2 - \rho_1}{\rho_1 + \rho_2} \right) - \frac{\rho_1 \rho_2}{(\rho_1 + \rho_2)^2} (U_1 - U_2)^2 \right)^{1/2}. \tag{7.25}$$

Show that there are the instabilities:
1. When $U_1 = U_2 = 0$, $\rho_1 > \rho_2$,   the growth rate of the Rayleigh-Taylor instability is

$$\gamma(k) = \pm \left(gk \left(\frac{\rho_1 - \rho_2}{\rho_1 + \rho_2} \right) \right)^{1/2}. \tag{7.26}$$

2. $g = 0$, $U_1 \neq U_2$,   the complex frequency, $\omega = \omega' + i\omega''$, of the Kelvin-Helmgoltz instability for $\rho_1 = \rho_2$ is

$$\omega(k) = k\left(\frac{U_1 + U_2}{2} \pm i\frac{U_1 - U_2}{2}\right). \tag{7.27}$$

◇

In the long-wavelength limit, when $k \to 0$, we see from equation (7.23) that $\omega(k) = k(gh)^{1/2}$. This dependence corresponds to the wave in the media without dispersion, with both the phase and group velocity equal $v_{\text{ph}} = v_{\text{g}} = (gh)^{1/2}$.

Performing integration of equation (7.2) and taking into account the mass conservation we find

$$\partial_t h + \partial_x(hv) = 0, \tag{7.28}$$

and

$$\partial_t v + v\partial_x v = -g\partial_x h. \tag{7.29}$$

Here the water depth, $h(x,t)$, is supposed to be a function of time and the x-coordinate, and $v(x,t)$ is the x-component of the fluid velocity.

In the linear on the perturbation amplitude approximation we recover the dependence $\omega(k) = k(gh)^{1/2}$. Equations (7.28) and (7.29) have a form of the gas dynamics equations for the gas with the specific heat magnitude equal 2. This set of equations can be solved with the hodograph transform. The Riemann wave can also be found. It describes the wave breaking on the shallow water surface.

7.1.2   The Rayleigh-Taylor Instability in the Thin Layer of Water

If one changes the sign of the right hand side in equation (7.29), the equation will describe the Rayleigh-Taylor instability of a thin layer of fluid in the shallow water approximation. Further we shall demonstrate that there is a broad variety of processes in unstable configurations that lead in the long-wavelength limit to similar equations. That is why to study the exact solutions to that set of equations is of generic interest.

We can write one of the solutions to equations (7.28) and (7.29) in the form

$$h = h_0(t) - h_2(t)x^2, \tag{7.30}$$

and

$$v = w(t)x. \tag{7.31}$$

This quite simple dependence of functions h and v on time and the coordinate have the meaning of the local approximation of the solution near singularity. Nevertheless, it is an exact solution to equations (7.28) and (7.29). This is an example of self-similar solutions. From equations (7.28), (7.29) and relationships given by (7.30) and (7.31) we find that functions $h_0(t)$, $h_2(t)$ and $w(t)$ obey ordinary differential equations

$$\dot{h}_0 + wh_0 = 0, \tag{7.32}$$

$$\dot{h}_2 + 3wh_2 = 0, \tag{7.33}$$

$$\dot{w} + w^2 = 2gh_2. \tag{7.34}$$

Changing the variable

$$w = \frac{\dot{s}}{s} \tag{7.35}$$

we express solutions to equations (7.32) and (7.33) in the form

$$h_0(t) = \frac{h_0(0)}{s}, \qquad h_2(t) = \frac{h_2(0)}{s^3}. \tag{7.36}$$

Then for function $s(t)$ we have equation

$$\ddot{s} = \frac{2gh_2(0)}{s^2} \tag{7.37}$$

with initial conditions: $s(0) = 1$, $\dot{s}(0) = w(0)$.

If $gh_2(0)$ is positive, we do not have any singular behavior of the solution. When $gh_2(0)$ is negative, it corresponds to the Rayleigh-Taylor instability, for $|w(0)|^2 < 4$ the solution $s(t)$ has a singularity: $s \to 0$ in a finite time equal to t_0. The solution describes the evolution of a parabolic in the x-direction water drop. We see that the drop, as follows from the expression for the water depth

$$h(t) = \frac{h_0(0)}{s}\left(1 - \frac{h_2(0)}{h_0(0)}\frac{x^2}{s^2}\right), \tag{7.38}$$

becomes more and more narrow: its width, $s(h_0(0)/h_2(0))^{1/2}$, tends to zero, meanwhile the height, $h_0(0)/s$, increases when $s \to 0$. In the vicinity of the singularity from equation (7.37) we find that

$$s(t) \approx (9gh_2(0)(t_0 - t)/2)^{2/3}. \tag{7.39}$$

We can also see from equations (7.31) and (7.35) that the function $s(t)$ is the Jacobian of transformation from the Euler coordinate x to the Lagrangian one x_0. In the Lagrange coordinates the velocity is equal to $v = \dot{s}x_0$, and the relationship between x and x_0 is given by the formula $x = s(t)x_0$.

7.1.3 The Rayleigh-Taylor Instability of a Compressible Shell.

As it has been demonstrated above (see equations (7.25,7.26)) an equilibrium with a heavy liquid upon a light one is unstable. The growth rate of the instability is

$$\gamma = \left(kg\frac{\rho_2 - \rho_1}{\rho_2 + \rho_1}\right)^{1/2}. \tag{7.40}$$

Now we discuss the problem of the Rayleigh-Taylor instability of a thin layer that admits an exact solution. We consider a thin layer with surface density σ, in the gravitational field $\mathbf{g} = -g\mathbf{e}_y$. The layer initially lies in the $y = 0$ plane, it separates the upper half-plane, $y > 0$, where the pressure of the massless gas is equal to p_1, and the lower half-plane, $y < 0$, where the pressure is

$$p_2 = p_1 - \sigma_0 g. \tag{7.41}$$

Here σ_0 is the initial value of the surface density. We consider a two-dimensional problem when all functions depend on two coordinates: x, y, and time t.

During the instability development the position, form and surface mass density of the layer change. The mass contained between two points whose coordinates at $t = 0$ are

$$(x = x_0, \ y = 0) \tag{7.42}$$

and

$$(x = x_0 + dx_0, \ y = 0), \tag{7.43}$$

is equal to $dm = \sigma_0 dx_0$. By virtue of the mass conservation we have

$$\sigma_0 dx_0 = \sigma dl, \tag{7.44}$$

where $dl = |d\mathbf{r}(t)|$ is a distance between the points (7.42) and (7.43) whose positions at time t are

$$\mathbf{r}(t) = (x_0 + \xi_x, \ \xi_y) \tag{7.45}$$

and

$$\mathbf{r}(t) + d\mathbf{r}(t) = \left(x_0 + \xi_x + \left(1 + \frac{\partial \xi_x}{\partial x_0} \right) dx_0, \ \xi_y + \frac{\partial \xi_y}{\partial x_0} dx_0 \right), \tag{7.46}$$

respectively. Here the components of the vector of the displacement, $\xi_x(x_0, t)$, $\xi_y(x_0, t)$, depend on x_0 and t. The vector of normal to the layer surface at the point $\mathbf{r}(t)$ is equal to $\mathbf{n} = (d\mathbf{r}(t) \times \mathbf{e}_z) / |d\mathbf{r}(t)|$.

Force acting on the layer element of the length dl is

$$d\mathbf{f} = -g\sigma dl\,\mathbf{e}_y + (p_1 - p_2)d\mathbf{r} \times \mathbf{e}_z. \tag{7.47}$$

Changing the Euler coordinates $\mathbf{r}(t)$ to the Lagrange coordinates $\mathbf{r}_0(t) = (x_0, 0)$, and using equations (7.41) and (7.44), we obtain expression for the force

$$d\mathbf{f} = -g\sigma_0 dx_0\,\mathbf{e}_y + (p_1 - p_2)dx_0 \frac{\partial \mathbf{r}}{\partial x_0} \times \mathbf{e}_z. \tag{7.48}$$

Calculating the Jacobian

$$\frac{\partial \mathbf{r}}{\partial x_0} = \left(\left(1 + \frac{\partial \xi_x}{\partial x_0} \right), \frac{\partial \xi_y}{\partial x_0} \right), \tag{7.49}$$

we obtain that

$$d\mathbf{f} = \left(g\sigma dl \frac{\partial \xi_y}{\partial x_0}, -g\sigma dl \frac{\partial \xi_x}{\partial x_0} \right). \tag{7.50}$$

Equations of motion give

$$\partial_{tt}\xi_x = g\partial_{x_0}\xi_y, \qquad \partial_{tt}\xi_y = -g\partial_{x_0}\xi_x. \tag{7.51}$$

We seek a solution to equations (7.51) in the form

$$\xi_x = \sum_k a_k \sin kx_0, \qquad \xi_y = \sum_k b_k \cos kx_0, \tag{7.52}$$

where a_k and b_k obey equations

$$\ddot{a}_k = -kgb_k, \qquad \ddot{b}_k = -kga_k. \tag{7.53}$$

We see that a_k and $b_k \sim \exp(i\omega t)$. For the frequency ω we have the dispersion equation

$$\omega^4 = k^2 g^2. \tag{7.54}$$

For the growth rate of the Rayleigh-Taylor instability it gives

$$\gamma = (kg)^{1/2}. \tag{7.55}$$

A particular solution to the problem can be written as

$$x = x_0 + a\exp((kg)^{1/2}t)\sin kx_0,$$

$$y = ya\exp((kg)^{1/2}t)\cos kx_0. \tag{7.56}$$

Exercise.   Demonstrate that the breaking occurs at $t_{\mathrm{br}} = (kg)^{-1/2}(\ln(1/ka))$. Demonstrate that the singularity in the (x,y) plane has the form of a cusp.

◇

From equations (7.51) it follows that the complex function $w(m,\tau) = x + iy$, where $m = x_0/g$, obeys equation

$$\partial_{\tau\tau}w = -i\partial_m w, \tag{7.57}$$

while for the complex conjugate function $w^*(m,\tau) = x - iy$ we have

$$\partial_{\tau\tau}w^* = i\partial_m w^*, \tag{7.58}$$

where w and w^* are considered as independent functions with $x = (w + w^*)/2$, and $y = -i(w - w^*)/2$.

The particular solution $w = im + \tau^2/2$ corresponds to a uniformly accelerated plane shell, while the solution

$$w(m,\tau) = x + iy = im^3 - i\frac{1}{4}m\tau^4 - \frac{1}{120}\tau^6 + \frac{3}{2}m^2\tau^2 \tag{7.59}$$

describes the local structure of the wave breaking and has the characteristic cubic dependence of the coordinate y on m.

For $w(m, \tau) \propto \exp(ikm)$ Eq.(7.57) describes exponentially growing and decaying modes for $k > 0$ and oscillatory modes with real frequency for $k < 0$. Particular solutions of such type are given by expressions (7.56). The intervals in k are interchanged in Eq.(7.58) for $w^*(m, \tau) \propto \exp(ikm)$.

For initial conditions $w_0(m) = w_1 \exp(im) + w_3 \exp(ikm)$ and $\partial_\tau w_0(m) = w_2 \exp(im) + w_4 \exp(ikm)$ and $w_0^*(m) = w_5 \exp(-im) + w_6 \exp(-ikm)$ and $\partial_\tau w_0^*(m) = w_7 \exp(-im) + w_8 \exp(-ikm)$ the solutions of Eqs.(7.57) and (7.58) are of the form

$$w(m, \tau) = (w_1 \cosh \tau + w_2 \sinh \tau) \exp(im) \tag{7.60}$$

$$+ (w_3 \cosh \sqrt{k}\tau + w_4 \sinh \sqrt{k}\tau) \exp(ikm),$$

and

$$w^*(m, \tau) = (w_5 \cos \tau + w_6 \sin \tau) \exp(-im) \tag{7.61}$$

$$+ (w_7 \cos \sqrt{k}\tau + w_8 \sin \sqrt{k}\tau) \exp(-ikm),$$

with a constant w_α.

We consider an initial configuration with k being an integer greater than one, $w_1 = w_2 = w_3 = 0$ and $w_4 = \kappa \ll 1$, and $w_5 = 1$, $w_6 = w_7 = w_8 = 0$. In this case Eqs.(7.60) and (7.61) describe a collapsing cylindrical shell, modulated with azimuthal number q and initial amplitude κ. As we can see in Fig. 7.2 (a). the radially expanding shell is unstable against perturbations of the form given by Eq.(7.60). After a finite time interval, compression wave breaking occurs with periodic azimuthal structure.

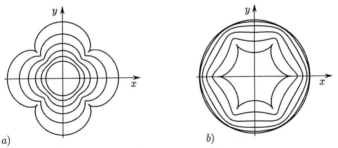

Fig. 7.2. Rayleigh-Taylor Instability of a converging (a) and a diverging (b) cylindrical shell.

A converging shell is also unstable, as we see in Fig. 7.2 (b). According to Eqs.(7.60) and (7.61), in the Lagrange coordinates we have a linear superposition of terms that describe the motion of the cylindrical shell and exponentially growing perturbations.

Now we consider the generic solution of Eqs.(7.57), (7.58). Application of the Lie group analysis shows that this equation admits 7 symmetry transformations represented by the operators:

$$X_\infty = w_1(m, \tau)\partial_w, \tag{7.62}$$

where $w_1(m, \tau)$ is a solution of Eq.(7.57),

$$X_1 = i\partial_m, \tag{7.63}$$

$$X_2 = \partial_\tau, \tag{7.64}$$

$$X_3 = 2m\partial_m + \tau\partial_\tau, \tag{7.65}$$

$$X_4 = 2im\partial_\tau - \tau w\partial_w, \tag{7.66}$$

$$X_5 = i\tau m\partial_\tau - im^2\partial_m - \frac{1}{4}(\tau^2 + 2im)w\partial_w, \tag{7.67}$$

$$X_6 = w\partial_w. \tag{7.68}$$

The operator X_∞ stems from the fact that Eq.(7.57) is linear with respect to $w(m, \tau)$ so that, to any solution $w(m, \tau)$, one can add any other solution $w_1(m, \tau)$. The operators X_1 to X_4 and X_6 correspond to the existence of stationary, uniform solutions and to the invariance with respect to stretching of the variables respectively. The operator X_5 represents the transformation

$$\bar{m} = \frac{m}{1 - am}, \qquad \bar{\tau} = \frac{\tau}{1 - am}, \qquad \bar{w} = (1 - am)^{1/2} \exp\left(\frac{ia\tau^2}{4(1 - am)}\right) w. \tag{7.69}$$

Under this transformation a solution $f(m, \tau)$ takes the form

$$w(m, \tau) = \frac{1}{(1 - am)^{1/2}} \exp\left(-\frac{ia\tau^2}{4(1 - am)}\right) f\left(\frac{m}{1 - am}, \frac{\tau}{1 - am}\right). \tag{7.70}$$

If we choose $f = i/(4\pi)^{1/2}$ and $a = -1/h$ in Eq.(7.70), and superpose $w = im + \tau^2/2$, we obtain a solution of the form

$$w(m, \tau) = im + \frac{\tau^2}{2} + \frac{w_0}{(4\pi(m + h))^{1/2}} \exp\left(i\frac{\tau^2}{4(m + h)}\right), \tag{7.71}$$

where $w_0 = i(h)^{1/2}$ is the initial perturbation amplitude and h is a complex parameter. The shell is initially a planar foil with perturbations localized in a region with size of order $|h|$. If h is imaginary and positive, $h = i|h|$, this solution describes perturbations that grow faster than exponential: $\propto \exp(\tau^2/4|h|)$. The typical singularity corresponds to the nonlinear superposition of compression and rarefaction waves. After a finite time the compression wave breaks, while it takes an infinite time for the rarefaction wave break to occur.

The superposition of solutions

$$w(m, \tau) = (w_1 \cosh\tau + w_2 \sinh\tau)\exp(im) + \frac{w_0}{(4\pi(m + h))^{1/2}} \exp\left(i\frac{\tau^2}{4(m + h)}\right) \tag{7.72}$$

describes the growth of nonlinear perturbations on a cylindrical shell. We see that both converging and diverging shells are unstable.

7.1.4 The Tearing Mode Instability of a Thin Current Sheet

Thin current layer of collisionless plasma is supposed to be in the $y = 0$ plane. We approximate the distribution of the electric current for $t = 0$ as $j(x, y) = I_0 \delta(y) = -n_0 e V L \delta(y)$. Here L is the current sheet thickness which is supposed to be much less than the electron Larmor radius, $r_{\mathrm{Be}} = m v_{\mathrm{Te}} c / e B_0$, $n_0 L$ is a surface density. The current sheet separates the upper- and lower half planes, where the magnetic field is equal to $B_0(y) = 2\pi I_0 \, \mathrm{sign}(y) \equiv B_0 \, \mathrm{sign}(y)$. The approximation of a thin current sheet is valid when

$$kL \ll \frac{m}{M}, \qquad \frac{V}{v_{\mathrm{Te}}} \gg \left(\frac{m}{M}\right)^{1/2}. \tag{7.73}$$

In the first expression k is a wave-number of perturbations when all values are supposed to be $\sim \exp(ikx)$.

Exercise. Demonstrate that the current sheet is unstable against perturbations $f(x, y, t) = f(y) exp(\gamma t + ikx)$ with the growth rate γ equal to

$$\gamma = \frac{kV}{(1 + \lambda_e |k|)^{1/2}} \left(\frac{m}{M}\right)^{1/2}. \tag{7.74}$$

Here $\lambda_e = mc^2 / 2\pi n_0 e^2 L$. The perturbations lead to the current filamentation, reconnection of the magnetic field lines and formation of a chain of magnetic islands. This is the tearing mode instability.

◇

We consider a quasineutral motion of a plasma in the x-direction. That is

$$n_e = n_i = n, \qquad v_{xe} = v_{xi} = v. \tag{7.75}$$

Since the initial configuration and perturbations are supposed to be independent of the z-coordinate, the z-component of the generalized momentum is conserved:

$$v_{ze} = V + \frac{e}{mc} A. \tag{7.76}$$

Here A is the z-component of the vector potential. The y-component of the magnetic field is expressed via A as $B_y = -\partial_x A$.

The two fluid hydrodynamic equations can be reduced in this approximation to describe the plasma motion along the $x-$axis. They take the form

$$\partial_t n + \partial_x (nv) = 0, \tag{7.77}$$

$$\partial_t v + v \partial_x v = -\frac{e}{Mc} \left(V + \frac{e}{mc} A\right) \partial_x A, \tag{7.78}$$

$$\partial_{xx} A + \partial_{yy} A = \frac{4\pi e}{c} \delta(y) n \left(V + \frac{e}{mc} A\right). \tag{7.79}$$

At the time $t = 0$ we have

$$A|_{|y| \to \infty} = B_0 |y| + o(1/y), \tag{7.80}$$

where $B_0 = 2\pi e n_0 LV/c$. We shall use this relationship as a boundary condition for $|y| \to \infty$. At the plane $y = 0$ we have the boundary condition

$$\partial_y A|_{y=+0} = \frac{\pi e L}{c} n \left(V + \frac{e}{mc} A|_{y=+0} \right),$$
(7.81)

that corresponds to the delta function in the right-hand side of equation (7.79).
We introduce the dimensionless variables

$$a = \frac{e}{V(mM)^{1/2}} (A - B_0|y|),$$
(7.82)

$$\nu = \frac{n}{n_0},$$
(7.83)

$$u = \frac{v}{V},$$
(7.84)

and the dimensionless coordinates

$$\tilde{x} = x/l, \qquad \tilde{y} = y/l, \qquad \tilde{t} = tV/l,$$
(7.85)

where l is a scalelength of perturbations. As a result equations (7.77—7.79) transform into a system

$$\partial_t \nu + \partial_x (\nu u) = 0,$$
(7.86)

$$\partial_t u + u \partial_x u = -\frac{1}{2} \partial_x (1 + a)^2,$$
(7.87)

with the Laplace equation for a function a in the region $|y| > 0$:

$$\partial_{xx} a + \partial_{yy} a = 0.$$
(7.88)

Here and below we omit the "tilde". The boundary conditions for the problem are

$$a|_{y \to \infty} = o(1/y),$$
(7.89)

and

$$\varepsilon \partial_y a|_{y \to 0} = \nu \left(1 + a|_{y \to 0} \right) - 1,$$
(7.90)

where the dimensionless parameter ε is equal to

$$\varepsilon = \frac{mc^2}{2\pi n_0 e^2 Ll} \equiv \frac{2\lambda_e^2}{Ll}.$$
(7.91)

Here $\lambda_e = c/\omega_{pe}$ is the collisionless skin depth. Further we suppose the parameter ε to be small compared to one, $\varepsilon \ll 1$. This means that the typical scale length of perturbations l is much larger than $2\lambda_e^2/L$. In this longwavelength limit we look for a solution to the problem in the form of the expansion in the parameter ε, when all functions can be represented as

$$f = f^{(0)} + \varepsilon f^{(1)} + \dots .$$
(7.92)

The zeroth order in the parameter ε gives a relationship between the vector potential $a^{(0)}$ and the plasma density $\nu^{(0)}$:

$$a^{(0)} = -1 + \frac{1}{\nu^{(0)}}. \tag{7.93}$$

Substituting this relationship into the right-hand side of the equation (7.87) we obtain a system of equations

$$\partial_t \nu + \partial_x (\nu u) = 0, \tag{7.94}$$

$$\partial_t u + u \partial_x u = -\frac{1}{\nu} \partial_x \left(\frac{1}{\nu} \right). \tag{7.95}$$

Here and below we have omitted the upperscript $^{(0)}$. The system has formally a form of the gas dynamics equations.

We perform a transformation from the Euler coordinates x, t to the Lagrange coordinates x_0, t' with

$$t = t', \qquad x = x_0 + \xi(x_0, t). \tag{7.96}$$

As we have discussed above there is a relationship between the displacement $\xi(x_0, t)$, velocity u and the density ν:

$$u = \partial_t \xi, \qquad \nu = \partial_x x_0 \equiv \frac{1}{1 + \partial_{x_0} \xi}. \tag{7.97}$$

Using these expressions in the Lagrange coordinates instead equations (7.94) and (7.95) we have equations

$$\partial_t \left(\frac{1}{\nu} \right) = \partial_{x_0} u, \tag{7.98}$$

and

$$\partial_t u = -\partial_{x_0} \left(\frac{1}{\nu} \right). \tag{7.99}$$

These equations are nothing more that the Cauchy-Riemann conditions for real and imaginary parts of an analytical function

$$w(\zeta) = 1 + \frac{1}{\nu(\zeta)} + i u(\zeta) \tag{7.100}$$

of the complex variable

$$\zeta = x_0 + it. \tag{7.101}$$

This analytical function is defined by its behavior on the real axis $t = 0$, that is by the initial conditions for the system (7.98)—(7.99): $\nu(x_0, 0)$ and $u(x_0, 0)$.

The displacement $\xi(x_0, t)$ as it follows from equations (7.97)—(7.99) obeys the Laplace equation

$$\partial_{tt} \xi + \partial_{x_0 x_0} \xi = 0. \tag{7.102}$$

As it is well known, the Cauchy problem for the Laplace equation is ill-posed. Nevertheless, studying the instability development from initial perturbations we need to analyze the Cauchy problem for equation (7.102). A solution to equation (7.102) is

$$\xi(x_0, t) = \Re\left\{ \int_0^{\zeta = x_0 + it} w(s) \, ds \right\}. \tag{7.103}$$

As a result we obtain that the density and velocity are equal to

$$\nu(x_0, t) = \frac{1}{1 + \Re\{w(\zeta)\}}, \tag{7.104}$$

and

$$u(x_0, t) = \Im\{w(\zeta)\}. \tag{7.105}$$

To demonstrate the typical behavior of the solution described by expressions (7.103)—(7.105) we consider two cases of initial conditions.
    A).

$$\nu(x_0, 0) = 1, \qquad u(x_0, 0) = \frac{\kappa x_0}{1 + x_0^2}. \tag{7.106}$$

Here κ is a constant. Analytical function given by expression (7.100) is equal to

$$w(\zeta) = i\frac{\kappa \zeta}{1 + \zeta^2}. \tag{7.107}$$

From equation (7.103) we find

$$\xi(x_0, t) = \frac{\kappa}{2} \arctan \frac{2x_0 t}{1 - t^2 + x_0^2}. \tag{7.108}$$

Calculating real and imaginary parts of the function $w(\zeta)$ we obtain

$$\nu(x_0, t) = \frac{(1 - t^2 + x_0^2)^2 + 4x_0^2 t^2}{(1 - t^2 + x_0^2)^2 + 4x_0^2 t^2 + \kappa \, t \, (1 - t^2 + x_0^2)}, \tag{7.109}$$

and

$$u(x_0, t) = \frac{\kappa \, x_0 (1 + t^2 + x_0^2)}{(1 - t^2 + x_0^2)^2 + 4x_0^2 t^2}. \tag{7.110}$$

According to these expressions at a finite time $t = 1$ the rarefaction wave break occurs with a hole formation in a plasma density distribution. The hole has a finite size equal to $\pi \kappa$.
    B).

$$\nu(x_0, 0) = 1, \qquad u(x_0, 0) = \kappa x_0 \exp(-x_0^2). \tag{7.111}$$

Here κ is a constant. The analytical function given by expression (7.100) is equal to

$$w(\zeta) = -i\kappa\zeta \exp(-\zeta^2). \tag{7.112}$$

From equation (7.103) we find

$$\xi(x_0, t) = \frac{\kappa}{2}\Re\{i\exp(-\zeta^2)\} = \frac{\kappa}{2}\exp(t^2 - x_0^2)\sin(2x_0 t) \tag{7.113}$$

Calculating real and imaginary parts of the function $w(\zeta)$ we obtain

$$\nu(x_0, t) = \frac{1}{1 - \kappa\exp(t^2 - x_0^2)(x_0\sin(2x_0 t) - t\cos(2x_0 t))} \tag{7.114}$$

and

$$u(x_0, t) = \kappa\exp(t^2 - x_0^2)(x_0\cos(2x_0 t) + t\sin(2x_0 t)). \tag{7.115}$$

We see that at a finite time $t = 1$ the compression wave break occurs with a formation of two singular regions where the plasma density tends to infinity.

7.1.5   Nonlinear Stage of the Instability of Electron Beam in a Plasma.

This instability develops when the electron beam with the density n_b propagates in a plasma in which the electron density is equal to n_p. The instability develops with an electron time scale. That is why we can assume that ions are at rest with a uniform density n_0. The electrons in the beam have the initial temperature equal to zero and move with respect to the plasma electrons with a velocity u_0. Equations that describe this two fluid system can be written as

$$\partial_t \nu_p + \partial_x(\nu_p v_p) = 0, \tag{7.116}$$

$$\partial_t v_p + v_p\partial_x v_p = -\frac{e}{m}E, \tag{7.117}$$

$$\partial_t \nu_b + \partial_x(\nu_b v_b) = 0, \tag{7.118}$$

$$\partial_t v_b + v_b\partial_x v_b = -\frac{e}{m}E, \tag{7.119}$$

$$\partial_x E = 4\pi e(n_0 - n_p - n_b), \tag{7.120}$$

where "p" and "b" stand for the plasma and the beam, respectively.

Linearizing this set of equations around an equilibrium solution, which is convenient to take in the frame where $v_p = u_p$, $v_b = 0$, we obtain the dispersion equation

$$\varepsilon(\omega, k) = 1 - \frac{\omega_{pe}^2}{(\omega - ku_0)^2} - \frac{\omega_{be}^2}{\omega^2} = 0. \tag{7.121}$$

Here $\omega_{pe} = (4\pi n_p e^2/m)^{1/2}$ and $\omega_{be} = (4\pi n_b e^2/m)^{1/2}$.

From the dispersion equation (7.121) it follows that the electron-plasma system is unstable when the wavenumber, k, is small enough: $k < \omega_{pe}/u_0$. The growth rate of the instability has a maximum for $k \approx \omega_{pe}/u_0$ which is equal to

$$\gamma_{max} \approx \omega_{pe}\left(\frac{n_b}{n_p}\right)^{1/3} \tag{7.122}$$

when $n_b/n_p \ll 1$. In the longwavelength limit, when $k \ll \omega_{pe}/u_0$, the growth rate is a linear function of the wavenumber

$$\gamma \simeq k u_0 \left(\frac{n_b}{n_p}\right)^{1/2}. \tag{7.123}$$

To obtain this expression we must write the dispersion equation (7.121) in the form

$$\varepsilon(\omega, k) \approx \frac{\omega_{pe}^2}{(ku_0)^2} + \frac{\omega_{be}^2}{\omega^2} = 0, \tag{7.124}$$

using approximation $\omega \ll ku_0$ and neglecting the first term equal unity, that is we neglect the term $\partial_x E$ in the left hand side of equation (7.120). This means that we use the approximation of quasineutrality

$$n_p + n_b = n_0. \tag{7.125}$$

From Eqs. (7.116), (7.118) and (7.125) we obtain that

$$e(n_p v_p + n_b v_b) = -I(t), \tag{7.126}$$

where $I(t)$ is a function of time, that is the electric current density does not depend on the x-coordinate. Eliminating the electric field from Eqs. (7.117), (7.119), and using equations (7.125) and (7.126), we obtain the system of equations

$$\partial_t \nu + \partial_x (\nu v) = 0, \tag{7.127}$$

$$\partial_t u + u \partial_x u = \partial_x \left(\frac{(1-u)^2 \nu}{1-\nu}\right). \tag{7.128}$$

Here we use dimensionless variables

$$\tilde{x} = \frac{x}{l}, \qquad \tilde{t} = \frac{tI}{en_0 u_0}, \qquad u = \frac{v_b e n_0}{I}, \qquad \nu = \frac{n_b}{n_0}, \tag{7.129}$$

with l being the scalelength of perturbations. Here and below we omit the "tilde".

Then we introduce new variables w, μ equal to

$$w = \ln(1 - u), \qquad \mu = \frac{1}{1 - \nu}. \tag{7.130}$$

The set of equations (7.127) and (7.128) is quasilinear. That is why we can apply the "hodograph" transformation. In variables (7.130) the "hodograph" transformation gives

$$\partial_\mu x = (2(\mu - 1)(1 - w) + w) \partial_\mu t - (1 - w)\partial_w t, \tag{7.131}$$

$$\partial_w x = \mu(\mu - 1)(1 - w)\partial_\mu t + w\partial_w t. \tag{7.132}$$

Eliminating x from these equations by cross-differentiation we find

$$\mu(\mu - 1)\partial_{\mu\mu} t - 2(\mu - 1)\partial_{\mu w} t + \partial_{ww} t + \partial_w t + 2\partial_\mu t = 0. \tag{7.133}$$

The generic solution to this equation can be expressed in terms of the integrals of the hypergeometric functions. Here we discuss only a particular solution to equation (7.133):

$$t = \frac{2}{1-u} + 2\ln(1-u) - \frac{\nu}{1-\nu}, \qquad (7.134)$$

$$x = 2u - \frac{2}{1-u} + 4\ln(1-u) - \frac{\nu u}{(1-\nu)^2}. \qquad (7.135)$$

These expressions describe the decay of the electron beam into two separated bunches.

We notice that we do not assume the smallness of the beam density, n_b, with respect to the plasma density, n_p. These densities in the longwavelength limit must satisfy equation (7.125).

In the limit $\nu \ll 1$ when the beam density is small with respect to the plasma density, the set of equations (7.127) and (7.128) reduces to

$$\partial_t \nu + \partial_x(\nu v) = 0, \qquad (7.136)$$

$$\partial_t u + u\partial_x u = \partial_x \nu. \qquad (7.137)$$

We will meet this set of equations while discussing other problems.

7.1.6   Some remarks on the behavior of unstable media

Above we have considered nonlinear regimes of several instabilities and firstly obtained the growth rates of those which depend linearly on the wavenumber. That is the dispersion equation has a form

$$\omega^2 = -k^2 c_b^2 \qquad (7.138)$$

with c_b being a constant of a dimension of velocity. Among these instabilities are the tearing mode instability (see equation (7.74) in the limit $|k|\lambda_e \ll 1$), the beam instability (see equation (7.123)), the elf-focusing, the Rayleigh-Taylor instability of "shallow" water and many other instabilities. Had there been the positive sign in the right-hand side of equation (7.138), it would have described the wave propagation in the nondispersive media. In our case the equation for the amplitude, a, takes the form

$$\partial_{tt} a + c_b^2 \partial_{xx} a = 0 \qquad (7.139)$$

as it follows from equation (7.138). This is the Laplace equation. The Cauchy problem for this equation is ill-posed as it is well known. However, studying the development of the unstable mode from initial perturbations we need to deal with the Cauchy problem for equation (7.139). We write the solution to equation (7.139) in terms of functions of complex variables. We introduce the coordinate $\zeta = x + ic_b t$. As a result we can write the solution to equation (7.139) in the form

$$a(x,t) = \Re\{f(\zeta)\} \equiv \Re\{f(x + ic_b t)\}$$

$$= \Re \left\{ a_0(x + ic_b t) + i \int_0^{x + ic_b t} \dot{a}_0(s) ds \right\}. \tag{7.140}$$

Here the initial data, $a_0(x)$ and $\dot{a}_0(x)$, are given on the real axis of the complex coordinate ζ. Expression (7.140) determines the real and imaginary parts of analytical function $f(\zeta)$.

In a generic case the analytical function $f(\zeta)$ has singularities in the upper half plane of complex variable $\zeta = x + ic_b t$. For example, if we have initial conditions

$$a_0(x) = \frac{1}{1 + x^2}, \qquad \dot{a}_0(x) = 0, \tag{7.141}$$

the function $f(\zeta)$ is equal to $f(\zeta) = 1/(1 + \zeta^2)$. It has poles in the points $\zeta = \pm i$. It means that we have a solution that describes a singularity reached in a finite time.

Arbitrary small perturbations of initial conditions can change the type and position of the singularitites. We can see that with $a_0(x) = 1/(1 + x^2) + \varepsilon/(0.25 + x^2)^2$, $\dot{a}_0(x) = 0$, where ε is supposed to be small. In this case the singularity described by $a(x, t) = \Re\{f(\zeta)\}$ is in the point $\zeta = \pm 0.5i$, while the function $f(\zeta)$ can be approximated locally here as $f(\zeta) \approx \varepsilon/(\zeta - 0.5i)^2$. This is the proof of the fact that the Cauchy problem for the elliptic equation (7.139) is ill-posed.

However, this conclusion is correct only in the case when the dispersion equation (7.138) is valid for the whole range of the wavenumbers: $0 < k^2 < \infty$. Really, the relationship (7.138) is correct just in the longwavelength limit when $k \to 0$; the growth rate, γ, is a linear function of k for $k \ll k_m$, where k_m is some wavenumber. Often the growth rate reaches its maximum value at $k \approx k_m$, and vanishes for $k \gg k_m$. In the case of the tearing mode instability $k_m = L^{-1}$, where L is the current sheet thickness; for the beam instability $k_m = \omega_{pe}/u_0$, etc. Further we shall take that fact into account to qualitatively describe the behavior of an unstable system beyond the singularity predicted in the longwavelength approximation.

We assume that in the vicinity of the singularity the function $f(\zeta)$ can be approximated as having a simple isolated pole in the point $\zeta_0 = ic_b t_0$:

$$f(\zeta) = \frac{1}{\zeta_0 - \zeta} = \frac{1}{ic_b(t_0 - t) - x}. \tag{7.142}$$

The Fourier transform of this function with respect to the coordinate x gives

$$f(k, t) = \int_{-\infty}^{+\infty} f(x, t) \exp(ikx) dx$$

$$= \int_{-\infty}^{+\infty} \frac{\exp(ikx)}{ic_b(t_0 - t) - x} dx = 2\pi i \chi(k) \exp(kc_b(t - t_0)). \tag{7.143}$$

Here $\chi(k)$ is the step function equal 1 for $k > 0$ and equal 0 for $k < 0$. The expression describes the exponential growth in time of perturbations with the growth rate $\gamma = kc_b$.

Since the growth rate beyond the longwavelength approximation should vanish for $k > k_m$, as we have noticed above, we put the function $f(k,t)$ to be equal to zero for $k > k_m$. The inverse Fourier transform then gives

$$\tilde{f}(x,t) = i \int_0^{k_m} \exp(-ikx + kc_b(t - t_0)) \, dk =$$

$$= -\frac{1 - \exp(-ik_m(x + ic_b(t - t_0)))}{x + c_b(t - t_0)}. \qquad (7.144)$$

At the initial stage until $k_m c_b(t - t_0) \gg 1$ this expression describes the dependence of the function $\tilde{f}(x,t)$ on the coordinates, which is similar to that obtained in the one-pole-approximation. The singularity starts to be formed at the point $x = 0$ for $t \to t_0$. In the vicinity of the singularity, when $|t_0 - t| \ll 1/k_m c_b$, the growth rate of the instability saturates. As a result the maximum value of the amplitude $a_{max}(0,t) \approx a(0,t_0)$ is proportional to k_m, and the modulations with the wavelength of order $2\pi/k_m$ appear. The function $\tilde{f}(\zeta)$, considered to be a function of the complex variable, is an entire function, that is, it has singularity at infinity in the ζ-plane.

Jean's Instability of Gravitating System

We consider the evolution of the small perturbations in the system which we describe by equations

$$\partial_t \rho + \nabla(\rho \mathbf{v}) = 0, \qquad (7.145)$$

$$\partial_t \mathbf{v} + (\mathbf{v}\nabla)\mathbf{v} = -\frac{1}{\rho}\nabla p - \nabla\varphi, \qquad (7.146)$$

$$\Delta\varphi = 4\pi G\rho. \qquad (7.147)$$

The first equation is the continuity equation, the second is the Euler equation with the force due to pressure gradient and the gravitation force in the right-hand side, and the third is the Poisson equation for the gravitation potential φ. Here G is the gravitation constant. The system may be a gravitating gas or stellar system.

Linearizing equations (7.145)- (7.147) around $\rho = \rho_0$, $\mathbf{v} = 0$, and $\varphi = 0$, we obtain

$$\partial_t \rho_1 + \rho_0 \nabla \cdot \mathbf{v}_1 = 0, \qquad (7.148)$$

$$\partial_t \mathbf{v}_1 = -\frac{c_s^2}{\rho_0}\nabla\rho_1 - \nabla\varphi_1, \qquad (7.149)$$

$$\Delta\varphi_1 = 4\pi G\rho_1, \qquad (7.150)$$

where $c_s^2 = \partial p / \partial \rho$. Writing all the variables in the form $f_1(\mathbf{x}, t) = f \exp(-i\omega t + i\mathbf{k} \cdot \mathbf{x})$ we obtain the dispersion equation for the frequency and the wave vector:

$$\omega^2 = k^2 c_s^2 - 4\pi G \rho_0. \tag{7.151}$$

We see that the perturbations with the wavelength larger than

$$\lambda_J = \frac{2\pi c_s}{(4\pi G \rho_0)^{1/2}} \tag{7.152}$$

are unstable ($\omega^2 < 0$ for $k < (4\pi G \rho_0 / c_s^2)^{1/2}$). The wavelength λ_J is called the Jean's length. We see that finite size system can be stable against gravitational contraction.

The condition for contracting of the gravitating cloud can also be written with the use of the virial theorem. Let us calculate the second derivative with respect to time of the momentum of inertia of particle ensembles. It reads

$$\frac{d^2 I}{dt^2} = \frac{d^2}{dt^2} \sum_\alpha m_\alpha \mathbf{x}_\alpha^2. \tag{7.153}$$

Differentiating the right-hand side of the equation two times we find

$$\frac{d^2 I}{dt^2} = \sum_\alpha 2 \left(m_\alpha \dot{\mathbf{x}}_\alpha^2 + \mathbf{x}_\alpha \cdot \mathbf{F}_\alpha \right), \tag{7.154}$$

where $\mathbf{F}_\alpha$ is the force acting on the particle α. We can rewrite equation (7.154) in the form

$$\frac{1}{2}\ddot{I} = 2K + U, \tag{7.155}$$

where K is the kinetic energy and U is the potential energy. The expression in the right-hand side of equation (7.155) is Clausius' virial (*Clausius, 1870*). If the second derivative of the momentum of inertia is positive, the cloud expands and it is stable against gravitational contraction. If $\ddot{I} = 0$ the cloud is in equilibrium. When $\ddot{I} < 0$ the system is unstable.

The kinetic energy in the cloud corresponds to the kinetic energy of turbulent motion $K = M v_t^2 / 2$, where M is the total mass of the cloud. The potential energy of a spherical homogeneous cloud with the radius R is $U = -5GM^2/3R$. It is stable if $v_t^2 > 5GM/3R$.

Now we consider the nonlinear regime of the gravitating cloud conraction. Assuming the spherical cloud to be cold ($p = 0$) we write equations (7.145)-(7.147) as

$$\partial_t \rho + v \partial_r \rho + \frac{2}{r} r v = 0, \tag{7.156}$$

$$\partial_t v + v \partial_r v = -g, \tag{7.157}$$

$$\frac{1}{r^2} \partial_r (r^2 g) = -4\pi G \rho. \tag{7.158}$$

Here $g = -\partial_r \varphi$ is the gravitation field. Changing the variables from the Euler r, t to the Lagrange r_0, t ones we find

$$\rho(r, t) = \rho_0(r_0)\frac{r_0^2}{r^2}\frac{\partial r_0}{\partial r}, \tag{7.159}$$

$$r^2 g = 4\pi G \int_0^r \rho r^2 dr \equiv GM(r), \tag{7.160}$$

where $M(r)$ is the mass inside a sphere with the radius r. From equation (7.159) it follows that the mass inside the Lagrange sphere does not depend on time and equals $M(r_0)$. If the density distribution inside the sphere is homogeneous, the mass is $M(r_0) = 4\pi\rho_0 r_0^3/3$. In the case the density distribution is given by $\rho = \rho_0(1 - r_0^2/R^2)$ the mass is $M(r_0) = 4\pi\rho_0(r_0^3/3 - r_0^5/5R^2)$. In the Lagrange coordinates the Euler equation takes the form

$$\ddot{r} = -\frac{GM(r_0)}{r^2}. \tag{7.161}$$

It describes the collapse that occurs in a finite time t_0 which is of the order of the free fall time $t_g = (4\pi G\rho_0)^{-1/2}$. Near the singularity as $t \to t_0$ we have

$$r(r_0, t) \approx \left(\frac{9}{2}GM(r_0)\right)^{1/3}(t_0 - t)^{2/3}, \qquad \rho(r, t) \approx \rho_0(r_0)(t_0 - t)^{-2}. \tag{7.162}$$

◇

Chapter 8

Systems Described by Nonlinear Elliptic Equations

Nonlinear elliptic equations. Plasma equilibrium in the magnetic field. The Grad-Shafranov equation. The Liouville equation and its solution with the conformal mapping. The Steward solution for vortex chains.

***

Here we shall discuss nonlinear elliptic equations of the form

$$\Delta u = F(u). \tag{8.1}$$

We restrict our consideration by the 2D case when the Laplace operator is $\Delta = \partial_{xx} + \partial_{yy}$.

Such an equation naturally appears to describe the plasma equilibrium in the magnetic field. Stationary 2D flows of the incompressible fluids are also described by equations of that type.

8.1 Plasma Equilibrium in the Magnetic Field

Now we derive the Grad-Shafranov equation to describe the plasma equilibrium.

At first we consider the case with a plane symmetry when both the poloidal magnetic field with components (B_x, B_y) and toroidal magnetic field, B_z , depend on the x, y−coordinates only. In this case the poloidal magnetic field can be expressed in terms of the z−component of the vector potential $A(x, y)$:

$$B_x = \partial_y A, \qquad B_y = -\partial_x A. \tag{8.2}$$

The force balance equations

$$\nabla p = \frac{1}{c}\mathbf{j} \times \mathbf{B}, \tag{8.3}$$

$$\mathbf{j} = \frac{c}{4\pi}\text{curl }\mathbf{B}, \tag{8.4}$$

$$\text{div }\mathbf{B} = 0 \tag{8.5}$$

in the 2D case can be rewritten in the form

$$\nabla p = \frac{1}{4\pi}\text{curl }\mathbf{B} \times \mathbf{B} = \frac{1}{4\pi}\left((\mathbf{B}\nabla)\mathbf{B} - \frac{1}{2}\nabla B^2\right) \equiv \frac{1}{4\pi}\left(\nabla A\Delta A - \frac{1}{2}\nabla B_z^2\right), \tag{8.6}$$

or

$$\nabla\left(p + \frac{1}{8\pi}B_z^2\right) = \frac{1}{4\pi}\nabla A\Delta A. \tag{8.7}$$

We introduce the function of the generalized pressure

$$\tilde{p} = p + \frac{1}{8\pi}B_z^2. \tag{8.8}$$

From equation (8.7) it follows that

$$\nabla\tilde{p} \times \nabla A = 0. \tag{8.9}$$

That is, the gradients of the generalized pressure and z–component of the vector potential are collinear. It means that the generalized pressure takes constant value on the surfaces $A =$ const and can be written as a function of A: $\tilde{p} = \tilde{p}(A)$. In this case from equation (8.7) we obtain that

$$\Delta A = \frac{d\tilde{p}}{dA}. \tag{8.10}$$

Now we consider an axially-symetric configuration. In polar coordinates r, z, φ it implies $\partial_\varphi = 0$. Here r, z are the poloidal coordinates while φ is the toroidal coordinate. We introduce the quantities

$$\psi(r, z) = 2\pi\int^r B_z r dr, \qquad I(r, z) = 2\pi\int^r j_z r dr. \tag{8.11}$$

From these expressions we find the poloidal components of the magnetic field and electric current:

$$B_r = -\frac{\partial_z\psi}{2\pi r}, \qquad B_z = \frac{\partial_r\psi}{2\pi r}, \qquad j_r = -\frac{\partial_z I}{2\pi r}, \qquad j_z = \frac{\partial_r I}{2\pi r}. \tag{8.12}$$

From equations of equilibrium (8.3), (8.4) it follows that $(\mathbf{B} \cdot \nabla)p = 0$ and $(\mathbf{j} \cdot \nabla)p = 0$, which means that the pressure gradient is zero along the magnetic field lines and along the electric current lines. In turn, it means that the magnetic field lines and the electric current lines lie on the surfaces $p(r, z) = const$

(magnetic surfaces) and ψ, I are constant on the magnetic surfaces. Therefore functions p and I can be expressed as the functions of ψ:

$$p = p(\psi), \qquad I = I(\psi). \tag{8.13}$$

The toroidal components of the magnetic field and electric current can also be expressed via ψ and I:

$$B_\varphi = \frac{2I}{cr}, \qquad j_\varphi = -\frac{c}{8\pi^2 r}\left(\partial_{rr}\psi - \frac{1}{r}\partial_r\psi + \partial_{zz}\psi\right). \tag{8.14}$$

Substituting these relationships into equation (8.3) we obtain the Grad-Shafranov equation. It reads

$$\partial_{rr}\psi - \frac{1}{r}\partial_r\psi + \partial_{zz}\psi = -16\pi^2 r^2\frac{dp}{d\psi} - \frac{8\pi^2}{c^2}\frac{dI^2}{d\psi}. \tag{8.15}$$

Specifying dependences $p = p(\psi), I = I(\psi)$, we find the equilibrium configuration. If we choose $dp/d\psi = const$ and $dI^2/d\psi = const$ we obtain the solution

$$\frac{\psi}{\psi_0} = \frac{1}{2}(bR^2 + r^2)z^2 + \frac{1}{8}(a-1)(r^2 - R^2)^2, \tag{8.16}$$

which describes a toroidal configuration with nested magnetic surfaces $\psi = const$. Here ψ_0, a, b, R are constants related to dependence of the pressure and electric current on ψ:

$$\frac{dp}{d\psi} = -\frac{a\psi_0}{16\pi^3}, \qquad \frac{dI}{d\psi} = -\frac{bR^2 c^2\psi_0}{8\pi^2}. \tag{8.17}$$

8.2 The Liouville Equation and its Solution with the Conformal Mapping

If we chose the generalized pressure, $\tilde{p}$, from equation (8.10) to be a function of the vector potential, A, of the form

$$\tilde{p}(A) = -\exp(A), \tag{8.18}$$

we obtain for A

$$\Delta A = -\exp(A). \tag{8.19}$$

This is the Liouville equation which we rewrite as

$$\Delta u = -\exp(u). \tag{8.20}$$

The solution to equation (8.20) can be found using the observation that under the conformal mapping $Z(\zeta)$, which transforms the complex variable $\zeta = x + iy$ into the variable $Z = X + iY$, the Laplace operator changes as

$$\Delta_\zeta = |Z'|^2\Delta_Z. \tag{8.21}$$

Here $\Delta_\zeta = \partial_{xx} + \partial_{yy}$, $\Delta_Z = \partial_{XX} + \partial_{YY}$, and $Z' = dZ/d\zeta$. If we consider the function $u(x, y)$ as a function of complex variable, the equation (8.20) changes undergo conformal mapping as

$$|Z'|^2 \Delta_Z u = -\exp(u). \tag{8.22}$$

Introducing a new function $v(x, y)$ equal

$$v = u - 2 \ln |Z'|, \tag{8.23}$$

we see that the function $v(X, Y)$ obeys equation

$$\Delta_Z v = -\exp(v) \tag{8.24}$$

which is also the Liouville equation in the X, Y-coordinates. That is, the formula (8.23) gives the relationship between different solutions.

Having found the solution $u(x, y)$ for the formula (8.23), we can find various solutions to the Liouville equation.

A particular solution to the Liouville equation can be found for example in the polar coordinates, r, φ, assuming that u is a function of the radius r only.

In this case we write the equation (8.20) as

$$\frac{1}{r}\frac{\partial}{\partial r} r \frac{\partial v}{\partial r} = -\exp(v). \tag{8.25}$$

We multiply the equation by r^2 and introduce a new variable and function:

$$l = \ln r, \qquad w = v + 2 \ln r. \tag{8.26}$$

In these variables, the equation (8.25) takes the form of the Liouville equation (8.20), when the dependent function depends on one Cartesian coordinate only

$$\frac{d^2 w}{dl^2} = -\exp(w). \tag{8.27}$$

The solution to this equation which satisfy the boundary conditions, $w(0) = 0$, $w'(0) = 0$, is

$$w(l) = -2 \ln \cosh l. \tag{8.28}$$

From this we have

$$v(r) = 2 \ln 2 - 2 \ln(r^2 + 1), \tag{8.29}$$

and

$$\exp(v) = \frac{4}{(r^2 + 1)^2}. \tag{8.30}$$

Using transformation (8.23) we obtain a generic solution to the Liouville equation:

$$\exp(u) \equiv G = \frac{4|Z'|^2}{(|Z|^2 + 1)^2}. \tag{8.31}$$

(See: *G.W.Walker, Proceed. Roy. Soc.* **A91**, *410 (1915)*; see also *T.Yeh, J. Plasma Physics,* **10**, *237 (1973))*.

If we choose the function $Z(\zeta)$ in the form

$$Z(\zeta) = \lambda + (1 + \lambda^2)^{1/2} \exp(i\nu\zeta), \qquad (8.32)$$

where λ and ν are parameters of the mapping, we find

$$G = \frac{\nu^2}{(\lambda \cosh \nu y - \nu^2 (1 + \lambda^2)^{1/2} \cos \nu x)^2}. \qquad (8.33)$$

This expression describes the current sheet in the plane y=0 with a street of magnetic islands as it is shown in Fig. 8.1

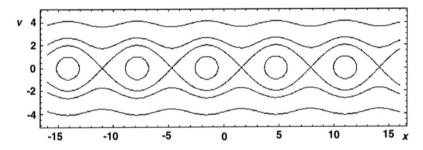

Fig. 8.1. Street of magnetic islands.

8.3 Vortices

Vortices. Plane flow of incompressible fluid. Interacting point vortices. Vortex collapse. Vortex dynamics in dispersive media. Stability of an infinite vortex chain. Bending instabilty of the vortex row. 3-D homogeneous collapse of generalized vorticity.

***

8.3.1 Plane Flow of Incompressible Fluid

Now we consider the plane flow of the incompressible fluid, when the flow depends on two cartesian coordinates, x and y. It can be described in terms of the stream function $\psi(x, y)$. The fluid velocity is

$$\mathbf{v} = \operatorname{curl}(\psi \mathbf{e}_z), \qquad (8.34)$$

and the vorticity has just the z-component equal to

$$\omega = \operatorname{curl}(\mathbf{v}) = -\Delta \psi. \qquad (8.35)$$

The Euler equations can be written as

$$\partial_t \omega = \{\psi, \omega\}, \tag{8.36}$$

or

$$\partial_t \Delta \psi = \{\psi, \Delta \psi\}. \tag{8.37}$$

Here the Poisson brackets are

$$\{f, g\} = \partial_x f \partial_y g - \partial_y f \partial_x g. \tag{8.38}$$

Equation (8.34), written in components, reads

$$\dot{x} = \partial_y \psi, \tag{8.39}$$

$$\dot{y} = -\partial_x \psi. \tag{8.40}$$

We see that two-dimensional fluid motion is described by the Hamiltonian equations with the Hamiltonian $\psi(x, y, t)$. These are also the characteristics of equation (8.36). The steady motion of the fluid is determined by equation

$$\{\psi, \Delta \psi\} = 0. \tag{8.41}$$

By virtue of relationship (8.38), which is nothing more than a Jacobian, we obtain that $\Delta \psi$ is a function of ψ. That is

$$\Delta \psi = F(\psi), \tag{8.42}$$

where $F(\psi)$ is an arbitrary function.

If we chose $F(\psi)$ to be equal

$$F(\psi) = -(1 - \varepsilon^2) \exp(2\psi) \tag{8.43}$$

we obtain the Liouville equation with solution

$$\psi = -\ln(\cosh \varepsilon y - \varepsilon \cos x). \tag{8.44}$$

This formula describes the vortex street known as the "Kelvin cat's eyes". It is formed at the nonlinear stage of the Kelvin-Helmholtz instability.

If we chose $F(\psi)$ to be equal

$$F(\psi) = -\frac{(1 - \varepsilon^2)}{2} \sinh(2\psi) \tag{8.45}$$

we obtain the sinh-Gordon equation

$$\Delta \psi = -\frac{(1 - \varepsilon^2)}{2} \sinh(2\psi) \tag{8.46}$$

with a solution

$$\psi = -\ln\left(\frac{\cosh \varepsilon y - \varepsilon \cos x}{\cosh \varepsilon y + \varepsilon \cos x}\right) = -2\operatorname{arc} \tanh\left(\frac{\varepsilon \cos x}{\cosh \varepsilon y}\right). \tag{8.47}$$

This expression describes the row of counter-rotating vortices.

8.3.2   Interacting Point Vortices

Assuming that we have a system of N point vortices with intensities κ_α ($\alpha = 1, \ldots, N$) localized at $\mathbf{x} = \mathbf{x}^\alpha(t)$ in an unbounded plane, we write equation (8.35) as

$$\Delta \psi = -\sum_\alpha \kappa_\alpha \delta(\mathbf{x} - \mathbf{x}^\alpha(t)). \tag{8.48}$$

It has a solution

$$\psi = -\frac{1}{2\pi} \sum_\alpha \kappa_\alpha \ln |\mathbf{x} - \mathbf{x}^\alpha(t)| \tag{8.49}$$

which is the Hamiltonian of a system (8.39)-(8.40). The Hamilton equations (8.39)-(8.40) take the form

$$\kappa_\alpha \dot{x}_i^\alpha = J_{ij} \frac{\partial \mathcal{H}}{\partial x_j^\alpha} = -\frac{1}{2\pi} J_{ij} \sum_{\beta \neq \alpha} \kappa_\alpha \kappa_\beta \frac{(x_i^\alpha - x_j^\beta)}{l_{\alpha\beta}^2}. \tag{8.50}$$

Here the antisymmetric matrix J_{ij} equals

$$J_{ij} = \begin{pmatrix} 0 & 1 \\ -1 & 0 \end{pmatrix}. \tag{8.51}$$

The Hamiltonian is given by

$$\mathcal{H} = -\frac{1}{2\pi} \sum_{\alpha < \beta} \kappa_\alpha \kappa_\beta \ln l_{\alpha\beta} = h, \tag{8.52}$$

where $l_{\alpha\beta} = |\mathbf{x}^\alpha - \mathbf{x}^\beta|$ is the distance between vortices. Written in the components, the Hamilton equations take the form

$$\dot{x}^\alpha = -\frac{1}{2\pi} \sum_{\beta \neq \alpha} \kappa_\beta \frac{y^\alpha - y^\beta}{(x^\alpha - x^\beta)^2 + (y^\alpha - y^\beta)^2}, \tag{8.53}$$

$$\dot{y}^\alpha = \frac{1}{2\pi} \sum_{\beta \neq \alpha} \kappa_\beta \frac{x^\alpha - x^\beta}{(x^\alpha - x^\beta)^2 + (y^\alpha - y^\beta)^2} \tag{8.54}$$

with $\alpha, \beta = 1, \ldots, N$.

The quantity $\mathcal{H}$ has the meaning of the energy of the interaction between vortices and is an integral of motion. Since the Hamiltonian is invariant with respect to translations along the coordinates, we have conserved quantities

$$Z_i = \sum_\alpha \kappa_\alpha x_i^\alpha = const. \tag{8.55}$$

Expressing $\mathcal{H}$ in terms of polar coordinates for the vortex position $(\rho^\alpha, \varphi^\alpha)$ we rewrite the Hamilton equations (8.50) as

$$\kappa_\alpha \rho^\alpha \dot{\rho}^\alpha = \frac{\partial \mathcal{H}}{\partial \varphi^\alpha}, \qquad \kappa_\alpha \rho^\alpha \dot{\varphi}^\alpha = -\frac{\partial \mathcal{H}}{\partial \rho^\alpha}. \tag{8.56}$$

From the rotation invariance of the Hamiltonian $\mathcal{H}$ it follows that

$$I = \sum_\alpha \kappa_\alpha (\rho^\alpha)^2 = const. \qquad (8.57)$$

Making use of equation (8.55) we rewrite equation (8.57) as

$$M = \sum_{\alpha,\beta} \kappa_\alpha \kappa_\beta l_{\alpha\beta}^2 = 2KI - 2Z_i^2. \qquad (8.58)$$

Here the sum of the vortex intensities

$$K = \sum_\alpha \kappa_\alpha \qquad (8.59)$$

is supposed to not vanish.

Multiplying (8.56) by ρ^α and sum over α we obtain

$$\sum_\alpha \kappa_\alpha (\rho^\alpha)^2 \dot{\varphi}^\alpha = -\frac{1}{2\pi} \sum_{\alpha<\beta} \kappa_\alpha \kappa_\beta \equiv V. \qquad (8.60)$$

The quantity V is called the virial of the vortex system. If the vortex system rotates rigidly, the angle velocity of rotation equals

$$\omega = V/I, \qquad (8.61)$$

where it is assumed that $I \neq 0$.

We also can see that the Hamilton equations (8.50) are invariant with respect to the transformation

$$x_i^\alpha \to \lambda x_i^{\alpha\prime}, \qquad t \to \lambda^2 t'. \qquad (8.62)$$

One vortex with the intensity κ remains at the same position. We assume that it is localized at $\mathbf{x} = 0$. The stream function is given by formula

$$\psi = -\frac{\kappa}{2\pi} \ln |\mathbf{x}|. \qquad (8.63)$$

The fluid velocity formed by the vortex has only the $\varphi-$ component and is equal to

$$v = \frac{\kappa}{2\pi |\mathbf{x}|}. \qquad (8.64)$$

Now we consider two vortices with intensities κ_1 and κ_2 localized at $\mathbf{x} = \mathbf{x}^1(t)$ and $\mathbf{x} = \mathbf{x}^2(t)$, respectively. We assume that the system barycenter is at the origin $\mathbf{x} = 0$. Then the velocity of motion of each vortex is equal to the velocity induced by the other vortex and directed perpendicular to the line $\mathbf{x}^1(t) - \mathbf{x}^2(t)$. It means that the velocities of the vortices are $v_1 = \kappa_2/2\pi l_{12}$ and $v_2 = \kappa_1/2\pi l_{12}$, where $l_{12} = |\mathbf{x}^1(t) - \mathbf{x}^2(t)| = const$. We find the angular velocity of the first vortex rotation by deviding v_1 by $|\mathbf{x}^1(t)|$, while the angular velocity of the rotation of the second vortex is $v_2/|\mathbf{x}^2(t)|$. Since they are equal to each other we have

$$|\mathbf{x}^1(t)| = \frac{\kappa_2}{\kappa_1 + \kappa_2} l_{12}, \qquad |\mathbf{x}^2(t)| = \frac{\kappa_1}{\kappa_1 + \kappa_2} l_{12}. \qquad (8.65)$$

The angular velocity of the rotation of the vortex pair is

$$\omega = \frac{\kappa_1 + \kappa_2}{2\pi l_{12}^2}. \tag{8.66}$$

If the intensity values of the vortices in the vortex pair are equal but intensities have opposite signs $\kappa_1 = -\kappa_2$, the barycenter coordinate is at infinity. The vortices move with equal velocity directed at the right angle to the line connecting the vortices. The velocity of the pair motion is equal to $v = \kappa_1/2\pi l_{12}$.

8.3.3   Vortex Collapse

In the case of homogeneous vortex collapse (E. G. Novikov, 1979) the distances between vortices have the same time dependence

$$l_{\alpha\beta}(t) = \lambda(t) l_{\alpha\beta}(t) \tag{8.67}$$

with $\lambda(0) = 1$ and $\lambda(t) \to 0$ as $t \to t_*$.

To have the collapsing motion, the virial of the vortex system (8.60) must vanish, $V = 0$. Since the value of the sum $L = \sum_\alpha \kappa_\alpha^2 > 0$, the intensity of the vortex system does not vanish. It means that the vortices do not annihilate during the collapse. The resulting vortex has the intensity higher than any of the vortices in the intitial configuration.

Conservation of the value of M given by equation (8.58) yields the second condition of the collapse $M = 0$.

In the polar coordinates in terms of inter-vortex distances we have

$$K(\rho^\alpha)^2 = \sum_\beta \kappa_\alpha l_{\alpha\beta}^2 - I. \tag{8.68}$$

Together with relationship

$$2\rho^\alpha \rho^\beta \cos(\varphi^\alpha - \varphi^\beta) = (\rho^\alpha)^2 + (\rho^\beta)^2 - l_{\alpha\beta}^2 \tag{8.69}$$

it implies that $I = 0$ and we have

$$\rho^\alpha(t) = \lambda(t) \rho^\alpha(0) \tag{8.70}$$

and

$$\varphi^\alpha(t) - \varphi^\beta(t) = \varphi^\alpha(0) - \varphi^\beta(0). \tag{8.71}$$

The motion is characterized by the scalar scaling factor $\lambda(t)$ and angular velocity $\omega(t)$. The necessary and sufficient conditions of the collapse motion are that the initial configuration satisfies the conditions

$$-\frac{2}{\kappa_\alpha(\rho^\alpha)^2} \frac{\partial \mathcal{H}}{\partial \varphi^\alpha}\bigg|_{t=0} = \frac{1}{t_*}, \qquad \frac{2}{\kappa_\alpha \rho^\alpha} \frac{\partial \mathcal{H}}{\partial \rho^\alpha}\bigg|_{t=0} = \omega(0). \tag{8.72}$$

Then from the Hamilton equations in the form (8.56) we obtain

$$\lambda(t) = \left(1 - \frac{t}{t_*}\right)^{1/2}, \qquad \text{and} \qquad \omega(t) = \frac{\omega(0)}{\lambda^2(t)}. \tag{8.73}$$

8.4   Vortex Dynamics in Dispersive Media

The Euler equations, in the framework of which we discussed the vortex dynamics, provide a model to describe the medium without dispersion. One of the simplest aproaches to study the vortices in dispersive media is to use the equations of so-called electron magnetohydrodynamics (EMHD). These equations have the form

$$\partial_t(\mathbf{B} - \Delta \mathbf{B}) = -\text{curl}\,(\text{curl}\,\mathbf{B} \times (\mathbf{B} - \Delta \mathbf{B})). \tag{8.74}$$

Here the space scale is chosen to be equal to the collisionless electron skin-depth, $d_e = c/\omega_{pe}$, and the time unit is $\omega_{Be}^{-1} = m_e c/eB$.

To derive the EMHD equations we have invoked the freezing of the z - component of the rotation of the generalized momentum $\text{curl}(\mathbf{p} - (e/c)\mathbf{A})$ into the electron fluid. Here $\mathbf{p}$ is the electron momentum and $\mathbf{A}$ is the vector potential. Since its motion is slow compared to the Langmuir time and its velocity is much smaller than speed of light c, the electron fluid can be regarded as incompressible. In addition, in this limit we can neglect the displacement curent in the Maxwell equations. Writing the electric current density as $\mathbf{j} = -en\mathbf{v}$, we obtain the relationship between the electron velocity and the magnetic field.We see that the electron component of the plasma moves with a mean velocity which depends on the magnetic field, $\mathbf{v} = \text{curl}\,\mathbf{B}$. In the nonrelativistic limit we write the generalized vorticity as $\mathbf{\Omega} = \mathbf{B} - \Delta \mathbf{B}$ and obtain equations (8.74). On the scales much larger than unity, one can neglect the term $\Delta \mathbf{B}$ in the expression for the generalized vorticity. In this limit the EMHD equations describe the freezing of the magnetic field into the electron motion. In the opposite limit the term $\Delta \mathbf{B}$ is dominant and, taking into account that $\Delta \mathbf{B} \sim \mathbf{v}$, we see that Eq. (8.74) becomes the Euler equation for the vorticity of the electron fluid. Such equations were discussed above.

In the linear approximation Eq. (8.74) describes the propagation of whistler waves. Their dispersion equation, that gives the relationship between the wave frequency and the wavevector, is

$$\omega = \frac{|\mathbf{k}|(\mathbf{k} \cdot \mathbf{B}_0)}{(1 + k^2)}. \tag{8.75}$$

In a two-dimensional case, taking $\mathbf{B}$ to be along the z-axis ($\mathbf{B} = B\mathbf{e}_z$), and assuming all the quantities to depend on the $x, y-$coordinates, we obtain from equations (8.74)

$$\partial_t(\Delta B - B) + \{B, (\Delta B - B)\} = 0. \tag{8.76}$$

Equation (8.76) is known as the Charney equation or the Hasegawa-Mima (HM) equation in the limit of zero drift velocity or the Electron-Magneto-Hydrodynamics equation. In this case linear perturbations with the dispersion equation (8.75) correspond to the Rossby waves, the drift waves or to the wistler waves, respectively.

Equation (8.76) has a discrete vortex solution for which the generalized vorticity is localized at the points $\mathbf{x} = \mathbf{x}^\alpha$:

$$\Omega = \Delta B - B = \sum_\alpha \kappa_\alpha \delta(\mathbf{x} - \mathbf{x}^\alpha(t)). \tag{8.77}$$

Then we have $B = \sum_\alpha B^\alpha$ with

$$B^\alpha(\mathbf{x}, \mathbf{x}^\alpha(t)) = -\frac{\kappa_\alpha}{2\pi} K_0(|\mathbf{x} - \mathbf{x}^\alpha(t)|). \tag{8.78}$$

Here and below $K_n(\xi)$ are modified Bessel functions. The curves $\mathbf{x}^\alpha(t)$ are determined by the characteristics of equation (8.76). The characteristic equations have the Hamiltonian form

$$\kappa_\alpha \dot{x}_i^\alpha = J_{ij} \frac{\partial \mathcal{H}}{\partial x_j^\alpha} = -\frac{1}{2\pi} J_{ij} \sum_{\beta \neq \alpha} \kappa_\alpha \kappa_\beta \frac{(x_i^\alpha - x_j^\beta)}{l_{\alpha\beta}^2} \tag{8.79}$$

similar to (8.50). The Hamiltonian is given by

$$\mathcal{H} = -\frac{1}{2\pi} \sum_{\alpha < \beta} \kappa_\alpha \kappa_\beta K_0(l_{\alpha\beta}) = h. \tag{8.80}$$

From these expressions the equations of motion of the vortices are

$$\dot{x}^\alpha = -\frac{1}{2\pi} \sum_{\alpha \neq \beta} \kappa_\beta \frac{y^\alpha - y^\beta}{(x^\alpha - x^\beta)^2 + (y^\alpha - y^\beta)^2} K_1\left(\left((x^\alpha - x^\beta)^2 + (y^\alpha - y^\beta)^2\right)^{1/2}\right), \tag{8.81}$$

$$\dot{y}^\alpha = \frac{1}{2\pi} \sum_{\alpha \neq \beta} \kappa_\beta \frac{x^\alpha - x^\beta}{(x^\alpha - x^\beta)^2 + (y^\alpha - y^\beta)^2} K_1\left(\left((x^\alpha - x^\beta)^2 + (y^\alpha - y^\beta)^2\right)^{1/2}\right). \tag{8.82}$$

In the case of the Euler hydrodynamics case, a point vortex is described by $(\kappa_\alpha/2\pi) \ln|\mathbf{x} - \mathbf{x}^\alpha(t)|$ (see (8.49)) instead of the expression (8.78) which involves the Bessel function $K_0(|\mathbf{x} - \mathbf{x}^\alpha(t)|)$. The later results in the shealding of the interaction between vortices at large distances. By direct inspection using asymptotic behaviour of the modified Bessel functions at large value of the argument, we may conclude that the velocity induced by one isolated vortex tends to zero at $|\mathbf{r}| \to \infty$ exponentially. The same is for the angular velocity of the vortex pair: it tends to zero exponentially as the distance between vortices increases.

Using an asymptotic representations of the modified Bessel functions

$$\begin{aligned} K_0(x) &\to -\ln x, \\ K_1(x) &\to 1/x, \\ K_2(x) &\to 2/x^2 + \ln x, \end{aligned} \tag{8.83}$$

as $x \to 0$, we may see that at short distances the behaviour of vortices in dispersive medium is the same as in the Euler hydrodynamics.

8.4.1   Stability of an Infinite Vortex Chain

Considering the problem of the stability of an infinite vortex chain we assume that all vortices have the same absolute intensity and take $|\kappa_\alpha| = 1$. In the initial equilibrium the vortices have coordinates

$$x^\alpha(0) = \alpha s, \quad y^\alpha(0) = 0, \quad -\infty < \alpha < +\infty \qquad (8.84)$$

and amplitudes $\kappa_\alpha = 1$. If the distance s between neighbouring vortices is much smaller than one , for $s \ll |y| \ll 1$ the chain separates two subregions, an upper and lower one, with opposite fluid velocity along x, $v_x = \mp U = \mp 1/(2s)$. This is equivalent to a vortex film with uniform surface density of generalized vorticity, $-1/s$. Far from the film, for $|y| \gg 1$, both B and v_x tend to zero exponentially.

In the analysis of a vortex chain stability we consider the motion of the α-th vortex with coordinates $x = \alpha s + x_1^\alpha(t)$ and $y = y_1^\alpha(t)$. Due to the translational invariance of the initial configuration we seek solutions of equations (8.79), linearized around the equilibrium configuration, of the form

$$x_1^\alpha = X \exp(\gamma t + i(\alpha\theta)), \qquad y_1^\alpha = X \exp(\gamma t + i(\alpha\theta)), \qquad (8.85)$$

with $0 < \theta < 2\pi$. If θ is small, the perturbation has the form of a sinusoidal wave with wavelength $\lambda = 2\pi/k = 2\pi s/\theta$, where k is the wavenumber. The perturbations grow exponentially in time, and the growth rate γ is given by

$$\gamma = \frac{1}{\pi}\left(\left(\sum_{\alpha=1}^{\infty}\frac{K_1(\alpha s)}{\alpha s}(1 - \cos\alpha\theta)\right)\left(\sum_{\beta=1}^{\infty}(K_2(\beta s) - \frac{K_1(\beta s)}{\beta s})(1 - \cos\beta\theta)\right)\right)^{1/2}.$$
$$(8.86)$$

If $s \ll 1$ and $\theta \gg 2\pi s$, $\lambda < 1$ and equation (8.86) reproduces the result obtained in (*H. Lamb, Hydrodynamics, 1932*):

$$\gamma \approx \frac{\theta(2\pi - \theta)}{4\pi s^2}. \qquad (8.87)$$

When $\theta \ll 1$, we have

$$\gamma \approx \frac{\theta}{2s^2} = kU, \qquad (8.88)$$

where $U = 1/(2s)$, which coincides with the growth rate of the Kelvin–Helmholtz instability. For $s \ll 1$ and $\theta < 2\pi s$, that is, for wavelengths greater than unity $(\lambda > 1)$, we have

$$\gamma \approx \frac{\theta^2}{\pi s^3} = \frac{k^2 U}{2\pi}. \qquad (8.89)$$

In the long wavelength limit the instability becomes slow compared to the Kelvin–Helmholtz instability. In the limit $s \gg 1$, when the distance between two neighboring vortices is larger than one, we obtain

$$\gamma \approx \frac{(1 - \cos\theta)\exp(-s)}{s\sqrt{2\pi}} \qquad (8.90)$$

and the instability is exponentially slow.

Bending Instabilty of the Vortex Row.

Let us consider a double chain of opposite vortices in which the coordinates and the amplitudes of point vortices are equal to

$$x^\alpha(0) = \alpha s + Ut, \quad y^\alpha(0) = \tfrac{1}{2}q, \quad -\infty < \alpha < +\infty, \quad \kappa_\alpha = -1 \qquad (8.91)$$

for the upper chain, and

$$x^\beta(0) = (\beta + \sigma)s + Ut, \quad y^\beta(0) = -\tfrac{1}{2}q, \quad -\infty < \beta < +\infty, \quad \kappa_\beta = 1 \qquad (8.92)$$

for the lower chain respectively. The distance between neighbour vortices in a chain is s, the distance between the chains in the y-direction is q, and the lower chain is shifted along the x-direction by σs: $\sigma = 0$ and $\sigma = 1/2$ correspond to the symmetrical and to the antisymmetrical configurations, respectively. Here

$$U = \frac{q}{\pi} \sum_{\beta=0}^{\infty} \frac{K_1(l_\beta)}{l_\beta}, \qquad l_\beta = ((\beta+\sigma)^2 s^2 + q^2)^{1/2} \qquad (8.93)$$

is the global velocity of the double chain in the x-direction. When $s \ll 1$ and $q \ll 1$:

$$U = \frac{1}{2s} \coth\left(\frac{\pi q}{s}\right) \qquad \text{for} \qquad \sigma = 0, \qquad (8.94)$$

and

$$U = \frac{1}{2s} \tanh\left(\frac{\pi q}{s}\right) \qquad \text{for} \qquad \sigma = \frac{1}{2}. \qquad (8.95)$$

Far from the vortex row the magnetic field and the electron fluid velocity tend to zero exponentially.

From equations (8.79) we can obtain, to the first order in perturbation amplitude, the linearized equation of motion of the vortices. Looking for solutions of the form

$$x_1^\alpha = X \exp(\gamma t + i(\alpha\theta)), \quad y_1^\alpha = X \exp(\gamma t + i(\alpha\theta)),$$
$$x_1^\beta = X' \exp(\gamma t + i(\beta\theta)), \quad y_1^\beta = Y' \exp(\gamma t + i(\beta\theta)), \qquad (8.96)$$

for the perturbations of the coordinates of vortices from the upper and the lower chain, respectively, we find the dispersion relation which gives the relationship between the real and imaginary parts of γ and the value of θ:

$$\det \begin{pmatrix} \gamma & C & B\exp(-i\sigma\theta) & D\exp(-i\sigma\theta) \\ E & \gamma & -F\exp(-i\sigma\theta) & -B\exp(-i\sigma\theta) \\ B\exp(i\sigma\theta) & -D\exp(i\sigma\theta) & \gamma & -C \\ F\exp(i\sigma\theta) & -B\exp(i\sigma\theta) & -E & \gamma \end{pmatrix} =$$

$$= (\gamma^2 - 2B\gamma + B^2 - (C+D)(E+F)) \times$$
$$\times (\gamma^2 + 2B\gamma + B^2 - (C-D)(E-F)) = 0. \qquad (8.97)$$

Here

$$B = i\frac{qs}{2\pi}\sum_{\alpha}\frac{K_2(l_\alpha)}{l_\alpha^2}(\alpha+\sigma)\sin(\alpha+\sigma)\theta,$$

$$C = \frac{1}{\pi}\sum_{\alpha}\frac{K_1(\alpha s)}{\alpha s}(1-\cos(\alpha\theta)) - \sum_{\alpha}\left(\frac{K_1(l_\alpha)}{l_\alpha} - \frac{q^2}{l_\alpha^2}K_2(l_\alpha)\right),$$

$$D = \sum_{\alpha}\left(\frac{K_1(l_\alpha)}{l_\alpha} - \frac{q^2}{l_\alpha^2}K_2(l_\alpha)\right)\cos(\alpha+\sigma)\theta,$$

$$E = \frac{1}{\pi}\sum_{\alpha}\frac{K_0(\alpha s)+K_2(\alpha s)}{2}(1-\cos(\alpha\theta)) +$$

$$+\sum_{\alpha}\left(\frac{K_1(l_\alpha)}{l_\alpha} - \frac{(\alpha+\sigma)^2 s^2}{l_\alpha^2}K_2(l_\alpha)\right),$$

$$F = \sum_{\alpha}\left(\frac{K_1(l_\alpha)}{l_\alpha} - \frac{(\alpha+\sigma)^2 s^2}{l_\alpha^2}K_2(l_\alpha)\right)\cos(\alpha+\sigma)\theta. \tag{8.98}$$

Solving this algebraic equation we obtain for γ:

$$\gamma_{1,2} = B \pm ((C+D)(E+F))^{1/2}.$$

$$\gamma_{3,4} = -B \pm ((C-D)(E-F))^{1/2}. \tag{8.99}$$

The symmetrical, $\sigma = 0$, vortex row is always unstable. In the limit $s \ll q \ll 1$ and $q \ll 2\pi s/\theta \ll 1$, we recover Rayleigh's result for the growth rate

$$\Re(\gamma) = \frac{\theta U(q\theta)^{1/2}}{s^{3/2}} = kU(kq)^{1/2} \tag{8.100}$$

of the bending instability of a finite width fluid stream. When the perturbation wavelength is larger than one and q ($q < 1 < 2\pi s/\theta$), we can estimate the instability growth rate as

$$\Re(\gamma) \approx k^2 U(kq)^{1/2}. \tag{8.101}$$

If the distance between neighbour vortices is larger than one, $s > 1$, the growth rate is exponentially small:

$$\Re(\gamma) \approx \frac{2\exp\left(-\frac{s}{2}\right)}{2\pi(2\pi s^3)^{1/4}}\left(\frac{K_1(q)}{q}\right)^{1/2}(1-\cos\theta)^{1/2}. \tag{8.102}$$

In the case of the antisymmetrical vortex row with $\sigma = 1/2$ we expect a more complicated behavior of the perturbations, compared to that of the symmetrical configuration. As noted in Lamb's monograph, in standard hydrodynamics the antisymmetrical von Karman's vortex row is stable for $q/s \approx 0.281$. We can see by direct inspection that for large distance between neighbour vortices the antisymmetric vortex row is stable for

$$3s^2/4 > q > s/2. \tag{8.103}$$

8.4.2 3-D Homogeneous Collapse of Generalized Vorticity.

Now we discuss the regime of nonlinear accumulation of generalized vorticity $\mathbf{\Omega} = \mathbf{B} - \Delta\mathbf{B}$ near the critical points of the magnetic and velocity fields. The critical point in the context of present discussion we define as the point where the frequency of small amplitude wave (8.75) vanishes. We see that the sufficient condition for the frequency to vanish at $\mathbf{x} = 0$ is that $\mathbf{B}_0(0) = 0$. Expanding the magnetic field in the vicinity of the critical point we find

$$B_i(\mathbf{x}) = A_{ij}x_j + A_{ijk}x_jx_k + \dots \quad . \tag{8.104}$$

Here the matrices of the field gradients are a 3×3 matrix A_{ij} and $3 \times 3 \times 3$ matrix A_{ijk}.

The formal solution of Eq. (8.74) is given by the formula obtained by Cauchy which we have discussed above:

$$B_i(\mathbf{x}, t) - \Delta B_i(\mathbf{x}, t) = \left(\frac{\partial x_i}{\partial x_j^0}\right)(B_j(\mathbf{x}^0, 0) - \Delta B_j(\mathbf{x}^0, 0)). \tag{8.105}$$

Here we use the fact that the Jacobian of the transformation from the Lagrange variables to the Euler coordinates $D = \det\left(\partial x_i/\partial x_j^0\right)$ is equal to unity due to incompressiblity of the electron fluid motion. The Euler and the Lagrange variables are related to each other by the formula $x_i = x_i^0 + \xi_i(\mathbf{x}^0, t)$, where $\xi_i(\mathbf{x}^0, t)$ is the displacement of the electron fluid element from its initial position x_i^0. From the Maxwell equations, $\mathbf{v} = -\mathrm{curl}\,\mathbf{B}$, under the condition $\mathbf{v} = \partial\xi/\partial t$, we obtain that the function $\xi(\mathbf{x}^0, t)$ obeys equation

$$\frac{\partial\xi_i}{\partial t} = -\varepsilon_{ijk}\left(\frac{\partial x_j^0}{\partial x_l}\right)\left(\frac{\partial B_l(\mathbf{x}^0, t)}{\partial x_k^0}\right), \tag{8.106}$$

where ε_{ijk} is the antisymmetric Ricci tensor.

We are seeking a solution of the problem in the framework of the self-similar solutions with homogeneous deformation. In this solution the deformation matrix $M_{ij} = \partial x_i/\partial x_j^0$ is assumed to depend only on time. This gives relationships between the Euler and Lagrange coordinates

$$x_i = M_{ij}(t)x_j^0 \tag{8.107}$$

as well as the expression for the velocity

$$\frac{\partial\xi_i}{\partial t} = w_{ij}(t)x_j, \tag{8.108}$$

where $w_{ij} = \dot{M}_{ik}M_{kj}^{-1}$.

In the self-similar solution the magnetic field space and time dependence is given by formula (8.104) with matrices of magnetic field gradients dependent on time. It reads

$$B_i(\mathbf{x}, t) = A_{ij}(t)x_j + A_{ijk}(t)x_jx_k, \tag{8.109}$$

while the generalized vorticity is given by

$$\Omega_i(\mathbf{x}, t) = A_{ij}(t)x_j + A_{ijk}(t)x_j x_k - A_{ikk}(t). \tag{8.110}$$

These expressions describe the vector field pattern in the vicinity of a null point of the third order. In the two-dimensional case it is the line of intersection of three separatrix surfaces. From equation (8.108) it follows that the fluid velocity is a linear function of the coordinates in agreement with expression (8.108). Taking into account the relationships given by equation (8.105) we obtain

$$A_{ij}(t) = M_{ik}(t)A_{kl}(0)M_{lj}^{-1}(t), \quad A_{ijk}(t) = M_{il}(t)A_{lmn}(0)M_{mj}^{-1}(t)M_{nk}^{-1}(t) , \tag{8.111}$$

while the deformation matrix $M_{mk}(t)$ acording to (8.106) obeys equation

$$\dot{M}_{ij} = -2\varepsilon_{ikl}M_{lm}M_{nk}^{-1}A_{mnj}(0). \tag{8.112}$$

In order to demonstrate the appearance of the generalized vorticity collapse we consider the tree-dimensional configuration which for $t = 0$ is described by the matrix $A_{ijk}(0)$ with

$$A_{123}(0) = A_{132}(0) = f_1, \quad A_{213}(0) = A_{231}(0) = f_2, \quad A_{312}(0) = A_{321}(0) = f_3, \tag{8.113}$$

with the remaining components equal to zero. From equations (8.112) and (8.111) it follows that for $f_1 < 0$, $f_2 > 0$, $f_3 = 0$, $f_1 + f_2 > 0$ the singularity is reached in a finite time interval

$$t_* = \frac{\ln\left(-\frac{f_2}{f_1}\right)}{4(f_1 + f_2)}. \tag{8.114}$$

When $t \to t_*$ we have

$$M_{11} \sim \frac{1}{(4|f_1|(t_* - t))^{1/2}} \to \infty, \qquad M_{33} \sim 4|f_1 f_2|^{1/2}(t_* - t) \to 0, \tag{8.115}$$

while for the nonzero components of the matrix A_{ijk} we find

$$A_{123} = A_{132} = -A_{213} = -A_{231} \sim -\frac{1}{4(t - t_*)}. \tag{8.116}$$

Chapter 9

Nonlinear Waves in Dissipative Media

Small amplitude wave propagation in dissipative media. Nonlinear waves in dissipative media. The Burgers equation. Shock waves. Weak shock wave. Strong shock wave. Hugoniot curves. The Cole-Hopf solution to the Burgers equation. Shock wave formation.

***

9.1 Small Amplitude Wave Propagation in Dissipative Media

Above we have discussed the wave propagation in the media without dissipation. Here we consider the effects of the viscosity and thermal conductivity on the acoustic wave propagation. We use a set of the gas-dynamics equations

$$\partial_t \rho + \operatorname{div}(\rho \mathbf{v}) = 0, \tag{9.1}$$

$$\partial_t \mathbf{v} + (\mathbf{v}\nabla)\mathbf{v} = -\frac{\nabla p}{\rho} + \mu \Delta \mathbf{v} + \left(\nu + \frac{\mu}{3}\right)\nabla \operatorname{div} \mathbf{v}, \tag{9.2}$$

$$p = p(\rho, T), \tag{9.3}$$

$$\partial_t s + (\mathbf{v}\nabla)s = \frac{\chi}{\rho T}\Delta T + \frac{\mu}{2T}\left(\frac{\partial v_i}{\partial x_k}\frac{\partial v_k}{\partial x_i} - \frac{2}{3}\delta_{ik}\frac{\partial v_j}{\partial x_j}\right)^2 + \frac{\mu}{3T}(\operatorname{div}\rho\mathbf{v})^2, \tag{9.4}$$

$$d\rho = (\partial_p \rho)_s \, dp + (\partial_s \rho)_p \, ds. \tag{9.5}$$

Here ρ is the density, $\mathbf{v}$ is the velocity, s is the entropy, T is temperature, p is the pressure, μ and ν are the coefficients of viscosity, χ is the coefficient of thermal conductivity.

Linearization of equations (9.1)—(9.5) around equilibria with $\rho = \rho_0$, $p = p_0$, $\mathbf{v}_0 = 0$, in the one-dimensional case gives

$$\partial_{tt} v - c_s^2 \partial_{xx} v = c_s^2 \tilde{\nu} \partial_{txx} v, \tag{9.6}$$

where

$$\tilde{\nu} = \frac{1}{c_s^2} \left(\left(\nu + \frac{4}{3}\mu \right) + \frac{\chi}{\rho_0} \left(\frac{1}{c_V} - \frac{1}{c_P} \right) \right), \tag{9.7}$$

c_V and c_P are the specific heat constants, and

$$c_s = \left(\left. \frac{\partial p}{\partial \rho} \right|_s \right)^{1/2} \tag{9.8}$$

is Newton's sound velocity.

The dispersion equation that comes from (9.6) has a form

$$k = \frac{\omega}{c_s} (1 - i\tilde{\nu}\omega)^{-1/2}. \tag{9.9}$$

That is the wavenumber has imaginary and real parts

$$k = k' + ik''. \tag{9.10}$$

In the limit $\tilde{\nu}\omega \approx k' c_s \tilde{\nu} \ll 1$ we have

$$k' \approx \omega/c_s \quad \text{and} \quad k'' \approx \tilde{\nu}\omega^2/2c_s. \tag{9.11}$$

That results in $v(x,t) \propto \exp(-i(\omega t - k'x) - k''x)$. This expression describes the sound wave decay due to viscosity and thermal conductivity.

Equation (9.6) can be transformed as

$$(\partial_t - c_s \partial_x)(\partial_t + c_s \partial_x) v = c_s^2 \tilde{\nu} \partial_{txx} v. \tag{9.12}$$

This is again a factorization of the wave operator.

We consider the wave that propagates to the right which is $v \approx v(t - x/c_s, \tilde{\nu}t)$. In the case when $\tilde{\nu}\omega \ll 1$ the dependence on $\tilde{\nu}t$ is slow compared to the dependence on $t - x/c_s$. For this, the wave $\partial_t \approx c_s \partial_x$, that is $\partial_t + c_s \partial_x \approx 2\partial_t$. Integrating ones on time equation (9.12) we obtain

$$(\partial_t + c_s \partial_x) v = \frac{c_s^2 \tilde{\nu}}{2} \partial_{xx} v. \tag{9.13}$$

In the frame that moves with the sound velocity c_s this equation reduces to the heat transport equation

$$\partial_t v = \frac{c_s^2 \tilde{\nu}}{2} \partial_{xx} v. \tag{9.14}$$

The solution to this equation is well known. As a result we write the solution to equation (9.13) that describes evolution of initially localized perturbation, $v(x,0) = \delta(x)$, as it follows

$$v(x,t) = \frac{1}{(\pi \tilde{\nu} c_s t)^{1/2}} \exp\left(-\frac{(c_s t - x)^2}{2\pi \tilde{\nu} c_s^2 t} \right). \tag{9.15}$$

9.2 The Burgers equation

Above, we derived the equation

$$\partial_t u + c_0 \partial_x u = \frac{\tilde{\nu}}{2} \partial_{xx} u, \tag{9.16}$$

which describes the wave propagation in dissipative media. This was a sound wave in a gas. The term in the right-hand side of the equation is due to the thermal conductivity and viscosity effects.

Nonlinear effects in the case without any dissipations were discussed above with the equation for the Riemann waves

$$\partial_t u + \left(c_0 + \frac{\gamma + 1}{2} u \right) \partial_x u = 0. \tag{9.17}$$

One can expect that the equation that takes into account both effects, dissipation and viscosity, should have the form

$$\partial_t u + u \partial_x u = \nu \partial_{xx} u. \tag{9.18}$$

It is really so. The equation (9.18) is called the Burgers equation.

Let us derive it from the equations of gasdynamics:

$$\partial_t \rho + \partial_x (\rho v) = 0, \tag{9.19}$$

$$\partial_t (\rho v) + \partial_x \left(\rho v^2 + p - \eta \partial_x v \right) = 0, \tag{9.20}$$

where for the pressure, p, and the density, ρ, we have the relationship

$$p = p_0 \left(\frac{\rho}{\rho_0} \right)^\gamma, \tag{9.21}$$

and η is viscosity.

We shall use the multiple scale expansion, changing independent variables:

$$\xi = \varepsilon^\beta (x - Vt), \tag{9.22}$$

$$\tau = \varepsilon^{2\beta} t. \tag{9.23}$$

Here the small parameter ε, is a measure of the role of dissipative effects (it means the longwavelength limit) and nonlinear effects (it means small amplitude approach). The velocity of the wave propagation, V, and the index β must be found during the solution of the system of the equations of gasdynamics (9.19) and (9.20).

We calculate the coordinate and time derivatives as

$$\partial_x = \varepsilon^\beta \partial_\xi, \tag{9.24}$$

$$\partial_t = \varepsilon^{2\beta} \partial_\tau - V \varepsilon^\beta \partial_\xi. \tag{9.25}$$

Then we expand the solution in the series

$$\rho = 1 + \varepsilon\rho^{(1)} + \varepsilon^2\rho^{(2)} + \cdots, \tag{9.26}$$

$$v = \varepsilon v^{(1)} + \varepsilon^2 v^{(2)} + \cdots, \tag{9.27}$$

$$\partial_x p = c_s^2(\rho)\partial_x \rho = \gamma\frac{p}{\rho}\partial_x \rho. \tag{9.28}$$

Substituting expansions (9.26)—(9.28) into equations (9.19) and (9.20) we obtain

$$\left(\varepsilon^{2\beta}\partial_\tau - V\varepsilon^\beta\partial_\xi\right)(1 + \varepsilon\rho^{(1)} + \varepsilon^2\rho^{(2)} + \cdots) +$$

$$+\varepsilon^{\beta+1}\partial_\xi((1 + \varepsilon\rho^{(1)} + \varepsilon^2\rho^{(2)} + \cdots)(v^{(1)} + \varepsilon v^{(2)} + \cdots)) = 0, \tag{9.29}$$

$$\left(\varepsilon^{2\beta+1}\partial_\tau - V\varepsilon^{\beta+1}\partial_\xi\right)((1 + \varepsilon\rho^{(1)} + \varepsilon^2\rho^{(2)} + \cdots)(v^{(1)} + \varepsilon v^{(2)} + \cdots)) +$$

$$+\varepsilon^\beta\partial_\xi\left((\varepsilon v^{(1)})^2 + \frac{c_s^2}{\gamma}(1 + \varepsilon\rho^{(1)} + \varepsilon^2\rho^{(2)} + \cdots)^\gamma -\right.$$

$$\left.-\eta\varepsilon^{\beta+1}\partial_\xi(v^{(1)} + \varepsilon v^{(2)} + \cdots)\right) = 0. \tag{9.30}$$

Then we write the terms of different order in ε. From equation (9.29) we have

$$\varepsilon^{\beta+1}\rfloor \qquad -V\partial_\xi\rho^{(1)} + \partial_\xi v^{(1)}, \tag{9.31}$$

$$\varepsilon^{2p+1}\rfloor \qquad \partial_\tau\rho^{(1)}, \tag{9.32}$$

$$\varepsilon^{\beta+2}\rfloor \qquad -V\partial_\xi\rho^{(2)} + \partial_\xi v^{(2)} + \partial_\xi\left(\rho^{(1)}v^{(1)}\right), \tag{9.33}$$

$$\varepsilon^{2\beta+2}\rfloor \qquad \partial_\tau\rho^{(2)}. \tag{9.34}$$

From equation (9.30) it follows that

$$\varepsilon^{\beta+1}\rfloor \qquad -V\partial_\xi v^{(1)} + c_s^2\partial_\xi\rho^{(1)}, \tag{9.35}$$

$$\varepsilon^{2\beta+1}\rfloor \qquad \partial_\tau v^{(1)} - \eta\partial_{\xi\xi}v^{(1)}, \tag{9.36}$$

$$\varepsilon^{\beta+2}\rfloor \qquad -V\partial_\xi(\rho^{(1)}v^{(1)} + v^{(2)}) + \partial_\xi(v^{(1)}v^{(1)}) +$$

$$+\frac{\gamma-1}{2}c_s^2\partial_\xi\left(\rho^{(1)}\rho^{(1)}\right) + c_s^2\partial_\xi\rho^{(2)}, \tag{9.37}$$

$$\varepsilon^{2\beta+2}\rfloor \qquad \partial_\tau v^{(2)} - \eta\partial_{\xi\xi}v^{(2)}. \tag{9.38}$$

Let the terms of lower order in ε be equal to zero (lowest order in ε is $\varepsilon^{2\beta+1}$); we obtain

$$-V\partial_\xi\rho^{(1)} + \partial_\xi v^{(1)} = 0, \tag{9.39}$$

$$-V\partial_\xi v^{(1)} + c_s^2\partial_\xi\rho^{(1)} = 0. \tag{9.40}$$

From these relationships we have

$$V^2 = c_s^2, \qquad v^{(1)} = V\rho^{(1)}. \tag{9.41}$$

In order for the effects of nonlinearity and dissipation to be comparable in their role, we need the index β to satisfy the algebraic equation

$$2\beta + 1 = \beta + 2, \qquad (9.42)$$

that is

$$\beta = 1. \qquad (9.43)$$

Taking that fact into consideration, we find that in the order ε^3

$$\partial_\tau \rho^{(1)} - V\partial_\xi \rho^{(2)} + \partial_\xi v^{(2)} + V\partial_\xi \left(\rho^{(1)}\rho^{(1)}\right) = 0, \qquad (9.44)$$

$$\partial_\tau \rho^{(1)} - V\partial_\xi \rho^{(2)} + \partial_\xi v^{(2)} + \eta\partial_{\xi\xi}\rho^{(1)} + (\gamma+1)\rho^{(1)}\partial_\xi\rho^{(1)} = 0, \qquad (9.45)$$

Eliminating $\left(-V\partial_\xi\rho^{(2)} + \partial_\xi v^{(2)}\right)$ from these equations, we obtain for $\rho^{(1)}$

$$\partial_\tau \rho^{(1)} + \frac{(\gamma+1)}{2}\rho^{(1)}\partial_\xi\rho^{(1)} = \frac{\eta}{2}\partial_{\xi\xi}\rho^{(1)}. \qquad (9.46)$$

To obtain exactly the Burgers equation in the form given by expression (9.18) we need to change the variables as it follows.

$$
\begin{aligned}
\tau &\rightarrow t, \\
\xi &\rightarrow x, \\
\tfrac{1}{2}\eta &\rightarrow \nu, \\
\tfrac{\gamma+1}{2}\rho^{(1)} &\rightarrow u.
\end{aligned}
\qquad (9.47)
$$

9.3   Stationary shock wave

9.3.1   Weak shock wave

We look for a solution to equation (9.18) in the form of a wave propagating along the x-axis with constant velocity V:

$$u(x,t) = u(x - Vt) \equiv u(X), \qquad (9.48)$$

where

$$X = x - Vt. \qquad (9.49)$$

Substituting expression (9.48) into equation (9.18), we find

$$\nu u'' - (u - V)u' = 0, \qquad (9.50)$$

where the prime denotes a derivative with respect to the X-coordinate.
Integrating equation (9.50) ones on X, we obtain

$$u' = \frac{(u - V)^2}{2\nu} + \text{const.} \qquad (9.51)$$

We use the boundary conditions for $X \rightarrow \pm\infty$:

$$u|_{X\rightarrow-\infty} = u_1, \qquad u|_{X\rightarrow+\infty} = u_2. \qquad (9.52)$$

With these boundary conditions we find that the constant in the right hand side of equation (9.51) equals $u_1 u_2/2$, and the wave speed is

$$V = \frac{u_1 + u_2}{2}. \tag{9.53}$$

Thus equation (9.51) can be rewritten as

$$u' = -\frac{1}{2\nu}(u - u_1)(u - u_2) \tag{9.54}$$

with solution

$$\frac{x - Vt}{\nu} = \frac{2}{u_1 - u_2} \ln \frac{u_1 - u}{u - u_2}, \tag{9.55}$$

or

$$u(x,t) = u_2 + \frac{u_1 - u_2}{1 + \exp\left(\dfrac{u_1 - u_2}{2\nu}(x - Vt)\right)}$$

$$= \frac{u_2 + u_1}{2} + \frac{u_1 - u_2}{2} \tanh\left(\frac{u_1 - u_2}{4\nu}(x - Vt)\right). \tag{9.56}$$

This solution describes a small amplitude shock wave. The shock wave amplitude is $u_1 - u_2$, it propagates with a constant velocity, which depends on the wave amplitude, $V = (u_2 + u_1)/2$. The width of the shock wave front is equal to

$$l = \frac{2\nu}{(u_1 - u_2)}. \tag{9.57}$$

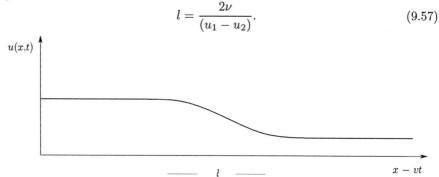

Fig. 9.1. Weak shock wave.

This relationship can be rewritten in the form of a dimensionless parameter

$$Re = \frac{l(u_1 - u_2)}{2\nu}, \tag{9.58}$$

which is the Reynolds number. At the shock wave front it equals one. The Reynolds number describes a relative role of dissipative and nonlinear effects. When $Re \ll 1$, the dissipation plays dominant role; when $Re \gg 1$, the nonlinear effects are more important. Since at the shock wave front $Re \approx 1$, the dissipative and nonlinear effects give a comparable contribution. That means the wave breaking here is stopped by the viscosity.

9.3.2 Strong Shock Wave. Hugoniot Curves

Now we consider the shock waves with arbitrarily high amplitude that obey equations (9.19), (9.20) to which we must add the energy equation

$$\partial_t \left(\rho \varepsilon + \frac{\rho u^2}{2} \right) = -\partial_x \left(\rho u \left(\varepsilon + \frac{u^2}{2} \right) + pu \right), \tag{9.59}$$

where ε is the specific internal energy. We consider the shock wave in the frame moving with the speed of shock front propagation and assume that the shock wave is at the coordinates $x = 0$. It separates regions (1) $x < 0$ and (2) $x > 0$ with homogeneous but different values ρ_1, u_1, ε_1 and ρ_2, u_2, ε_2.

We consider the discontinuity at $x = 0$ as a thin layer containing large gradients of all parameters. To find the boundary conditions at the discontinuity we integrate equations (9.19), (9.20) and (9.59) from $-\delta$ to δ as in the case

$$\int_{-\delta}^{\delta} \partial_t (\rho u) dx = -\int_{-\delta}^{\delta} \partial_x (p + \rho u^2) dx. \tag{9.60}$$

In the limit $\delta \to 0$ the integrals on the left hand side are proportional to δ, and vanish as $\delta \to 0$. That corresponds to the absence of mass, momentum and energy accumulation at the shock wave front. The integrals on the right-hand side give the relationship between the fluxes on each side of the discontinuity:

$$\rho_1 u_1 = \rho_2 u_2, \tag{9.61}$$

$$p_1 + \rho_1 u_1^2 = p_2 + \rho_2 u_2^2, \tag{9.62}$$

$$\varepsilon_1 + \frac{p_1}{\rho_1} + \frac{u_1^2}{2} = \varepsilon_2 + \frac{p_2}{\rho_2} + \frac{u_2^2}{2}. \tag{9.63}$$

The last equation can be rewritten as

$$h_1 + \frac{u_1^2}{2} = h_2 + \frac{u_2^2}{2}, \tag{9.64}$$

where $h = \varepsilon + p/\rho$ is the specific enthalpy. These relationships do not contain the dissipation terms due to viscosity and thermal conductivity since they are related to the regions where all the values are homogeneous. However inside a thin layer in the vicinity $x = 0$ where the flow variables change exceedingly sharply, but continuously, the dissipation effects must be taken into account.

Equations (9.61)-(9.63) form a system of three algebraic equations for six variables ρ_1, u_1, ε_1 and ρ_2, u_2, ε_2. We have assumed that the functions $\varepsilon(p, \rho)$ or $h(p, \rho)$ are known. Knowing the values of ρ_2, p_2 ahead of the discontinuity and assuming the pressure behind the shock front p_1 we can calculate all the remaining variables. In place of density we introduce the specific volumes

$$V_1 = 1/\rho_1 \quad \text{and} \quad V_2 = 1/\rho_2. \tag{9.65}$$

From (9.61)-(9.63) we have

$$\frac{\mathcal{V}_1}{\mathcal{V}_2} = \frac{u_1}{u_2}, \tag{9.66}$$

$$u_1 - u_2 = ((p_2 - p_1)(\mathcal{V}_1 - \mathcal{V}_2))^{1/2}, \tag{9.67}$$

$$\varepsilon_2(p_2, \mathcal{V}_2) - \varepsilon_1(p_1, \mathcal{V}_1) = \frac{1}{2}(p_2 + p_1)(\mathcal{V}_1 - \mathcal{V}_2). \tag{9.68}$$

By analogy with the equation relating initial and final pressures and volumes during the adiabatic compression of a gas, the equation (9.68) is called the shock adiabatic or the Hugoniot curve. It is represented by the function $p_2 = H(\mathcal{V}_2, p_1, \mathcal{V}_1)$. The Hugoniot curve differs from ordinary isentropres,which belong to a one-parametric family of curves $p = P(\mathcal{V}, s)$, parametrized only by the entropy s. The Hugoniot curve is parametrized by two parameters, the initial pressure p_2 and volume $\mathcal{V}_2$ in the region ahead the shock wave front.

The shock wave equations for a perfect gas with constant specific heats can be solved explicitly. Substituting the relationships for specific energy and enthalpy

$$\varepsilon = c_v T = \frac{1}{\gamma - 1}p\mathcal{V}, \qquad h = c_p T = \frac{1}{\gamma - 1}p\mathcal{V} \tag{9.69}$$

into equation (9.68) we obtain the expression for the Hugoniot curve

$$\frac{p_2}{p_1} = \frac{(\gamma + 1)\mathcal{V}_1 - (\gamma - 1)\mathcal{V}_2}{(\gamma + 1)\mathcal{V}_2 - (\gamma - 1)\mathcal{V}_1}, \tag{9.70}$$

from which the ratio of the specific volumes is

$$\frac{\mathcal{V}_2}{\mathcal{V}_1} = \frac{(\gamma - 1)p_2 + (\gamma + 1)p_1}{(\gamma + 1)p_2 + (\gamma - 1)p_1}. \tag{9.71}$$

The Hugoniot curve (9.70) is shown in Fig. 9.2.

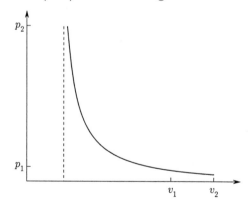

Fig. 9.2. Hugoniot curve.

From equation (9.71) we see that the density ratio across a very strong shock wave with $p_2 \gg p_1$ does not increase unlimitedly, but approaches a certain value, which is equal to

$$\frac{\rho_2}{\rho_1} = \frac{V_1}{V_2} = \frac{\gamma + 1}{\gamma - 1}. \tag{9.72}$$

For monoatomic gas with $\gamma = 5/3$ it is equal to 4.

From equations (9.61)-(9.63) we find the gas velocities with respect to the shock wave front

$$u_1^2 = V_1 \frac{p_2 - p_1}{V_1 - V_2}, \qquad u_2^2 = V_2 \frac{p_2 - p_1}{V_1 - V_2}. \tag{9.73}$$

Comparing the gas velocities with corresponding speeds of sound $c_s^2 = (\partial p/\partial \rho)_s = \gamma p/\rho = \gamma p V$, we see

$$\left(\frac{u_1}{c_{s,1}}\right)^2 = \frac{(\gamma - 1) + (\gamma + 1)p_2/p_1}{2\gamma}, \tag{9.74}$$

$$\left(\frac{u_2}{c_{s,2}}\right)^2 = \frac{(\gamma - 1) + (\gamma + 1)p_1/p_2}{2\gamma}. \tag{9.75}$$

These equations show that the gas flows into discontinuity with supersonic velocity $u_1 > c_{s,1}$ and flows out with subsonic velocity $u_2 > c_{s,2}$, and the shock wave propagates at a subsonic velocity with respect to compressed gas behind it and with supersonic velocity with respect to nondisturbed gas ahead of it. The velocity of the shock wave propagation is u_1. In the weak shock wave with $p_2 = p_1 + \delta p$, $\delta p \ll p$, $p = (p_1 + p_2)/2$, we have

$$u_1 \approx c_{s,1}\left(1 + \frac{\gamma + 1}{4\gamma}\frac{\delta p}{p}\right), \qquad u_2 \approx c_{s,2}\left(1 - \frac{\gamma + 1}{4\gamma}\frac{\delta p}{p}\right). \tag{9.76}$$

Velocity of propagation of the strong shock wave u_1 increases with p_2/p_1; on the other hand for $p_2 \gg p_1$ the ratio $u_2/c_{s,2} \to ((\gamma - 1)/2\gamma)^{1/2} < 1$.

Calculating the entropy $s = c_v \ln p V^\gamma$ we obtain that the difference between the entropy on each side of the shock front is

$$s_2 - s_1 = c_v \ln \frac{p_2 V_2^\gamma}{p_1 V_1^\gamma} = c_v \ln \left(\frac{p_2}{p_1}\left(\frac{(\gamma - 1)p_2 + (\gamma + 1)p_1}{(\gamma + 1)p_2 + (\gamma - 1)p_1}\right)^\gamma\right). \tag{9.77}$$

We see that across a compression shock wave the entropy increases which indicates that irreversible processes occur in the shock front and that in a perfect gas there are just compression shock waves.

9.3.3 A Solution to the Burgers Equation with the Cole-Hopf Transformation.

A remarkable property of the Burgers equation is the conservation of the integral

$$\int_{-\infty}^{+\infty} u(x,t)dx = S, \tag{9.78}$$

where $u(x, t)$ is a solution to the Burgers equation. That is, if $u(x, t)$ vanishes at infinity, where $|x| \to \infty$, the surface under the curve $u(x, t)$ remains constant. To prove that, we calculate the time derivative of S:

$$\partial_t S = \int_{-\infty}^{+\infty} \partial_t u \, dx = \int_{-\infty}^{+\infty} (-u\partial_x u + \nu \partial_{xx} u) \, dx =$$

$$= \int_{-\infty}^{+\infty} \partial_x \left(\nu \partial_x u - \frac{u^2}{2} \right) dx = \nu \partial_x u - \frac{u^2}{2} \Big|_{-\infty}^{+\infty} = 0. \tag{9.79}$$

Conservation of the value of S helps us to analyze the asymptotic behavior, when $t \to \infty$, of solutions to the Burgers equation. For example, if initial distribution of $u(x, 0)$ has a form of localized pulse (see Fig. 9.3 a), then the nonlinear effects result in the leading edge steepening and the pulse amplitude decreasing caused by the dissipation. Asymptotically for $t \to \infty$, the pulse takes a form of a triangular:

$$u(x, t) = \begin{cases} x/t, & 0 < x < b(t), \\ 0, & x < 0, \ x > b(t). \end{cases} \tag{9.80}$$

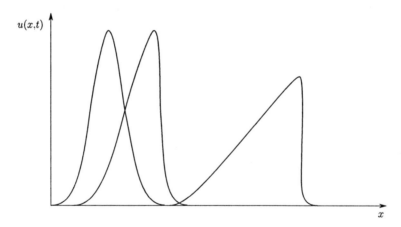

Fig. 9.3. Nonlinear evolution of the pulse.

The triangular height, $u(b(t), t) = b(t)/t = h(t)$, and its width, $b(t)$, can be found from the conservation of the area of the triangular surface:

$$S = \frac{h(t)b(t)}{2} = \frac{b^2(t)}{2t}, \tag{9.81}$$

since $h(t) = b(t)/t$. As a result we have

$$b(t) = (2St)^{1/2}, \qquad h(t) = \left(\frac{2S}{t} \right)^{1/2}. \tag{9.82}$$

Hopf and Cole showed that the Burgers equation can be transformed into the diffusion equation (the thermal conductivity equation). We develop the Cole-Hopf transformation in two steps. First, we introduce the function $\phi(x,t)$ defined as

$$u(x,t) = -\partial_x \phi. \tag{9.83}$$

We find the differential equation that $\phi(x,t)$ obeys from equations (9.18) and (9.83). It is

$$-\partial_x \left(\partial_t \phi - \frac{1}{2} (\partial_x \phi)^2 - \partial_{xx} \phi \right) = 0. \tag{9.84}$$

Here and below we use the variables normalized to have $\nu = 1$. Integrating this equation with respect to x, we obtain

$$\partial_t \phi - \frac{1}{2} (\partial_x \phi)^2 - \partial_{xx} \phi = f(t). \tag{9.85}$$

The function of integration, $f(t)$, can be eliminated by introducing a new dependent variable $\psi(x,t)$ defined by

$$\phi(x,t) = \psi(x,t) + \int_0^t f(s)ds. \tag{9.86}$$

Then, as a second step, we introduce a function $z(x,t)$ which is

$$z(x,t) = \exp \left(\frac{\psi(x,t)}{2} \right). \tag{9.87}$$

Substituting the function $z(x,t)$ into equation (9.85), we obtain

$$\partial_t \psi - \frac{1}{2} (\partial_x \psi)^2 - \partial_{xx} \psi = \frac{2}{z} \partial_t z - \frac{2}{z^2} (\partial_x z)^2 -$$

$$-\frac{2}{z^2} \left(z \partial_{xx} z - (\partial_x z)^2 \right) = \frac{2}{z} (\partial_t z - \partial_{xx} z) = 0. \tag{9.88}$$

Equation (9.88) states that $z(x,t)$ satisfies a linear diffusion equation

$$\partial_t z - \partial_{xx} z = 0. \tag{9.89}$$

The initial-value problem for the Burgers equation has a solution satisfying equation (17.18) with a function $u(x,t)$ at time $t = 0$ specified by a function u_0 :

$$u(x,0) = u_0(x). \tag{9.90}$$

The initial condition for a function $z(x,t)$ is

$$z(x,0) = z_0(x), \tag{9.91}$$

where

$$z_0(x) = \exp\left(\frac{\psi_0(x)}{2}\right) = \exp\left(-\frac{1}{2}\int_{-\infty}^{x} u_0(x')dx'\right) \tag{9.92}$$

The diffusion equation (9.89) is linear and it can be solved by one of several well known methods. If $z(x,t)$ is a solution to equation (9.89) with initial condition (9.92), the function $u(x,t)$ is obtained as

$$u(x,t) = -2\partial_x \ln(z(x,t)) = -\frac{2}{z(x,t)}\partial_x z(x,t). \tag{9.93}$$

The solution of the initial-value problem for a diffusion equation is given by expression

$$z(x,t) = \frac{1}{(2\pi t)^{1/2}} \int_{-\infty}^{+\infty} z_0(x')\exp\left(-\frac{(x-x')^2}{4t}\right)dx'. \tag{9.94}$$

From equations (9.93) and (9.94) we find the function $u(x,t)$ which is given by formula

$$u(x,t) = \frac{\displaystyle\int_{-\infty}^{+\infty} \frac{x-x'}{t}\exp\left(-\frac{(x-x')^2}{4t} - \frac{1}{2}\int_{-\infty}^{x'} u_0(x'')dx''\right)dx'}{\displaystyle\int_{-\infty}^{+\infty}\exp\left(-\frac{(x-x')^2}{4t} - \frac{1}{2}\int_{-\infty}^{x'} u_0(x'')dx''\right)dx'}. \tag{9.95}$$

In the nominator and denominator, the main contribution to the integral value comes from the neighborhoods of extremum points. Stationary points can be found by the differentiation of

$$-\frac{(x-x')^2}{4t} - \frac{1}{2}\int_{-\infty}^{x'} u_0(x'')dx''$$

with respect to x'. As a result we obtain equation

$$\frac{\partial}{\partial x'}\left(-\frac{(x-x')^2}{4t} - \frac{1}{2}\int_{-\infty}^{x'} u_0(x'')dx''\right)\Bigg|_{x_0} = \frac{1}{2}\left(\frac{x-x_0}{t} - u_0(x_0)\right) = 0. \tag{9.96}$$

This is an equation for x_0. From this it follows that

$$x = x_0 + u_0(x_0)t. \tag{9.97}$$

Substituting these dependencies into equation (9.95), we obtain that

$$u(x,t) = u_0(x_0). \tag{9.98}$$

The relationships given by expressions (9.97) and (9.98) define a solution to equation

$$\partial_t u + u \partial_x u = 0, \tag{9.99}$$

written in the Lagrange coordinates. Here the stationary point x_0 is the Lagrange coordinate. Equation (9.99) approximates the Burgers equation (9.18) in the limit when the dissipation effects are negligible. As we know, the obtained dependency describes the wave front steepening and the tendency to form the shock wave.

We know that the solution (9.97) and (9.98) is valid until the wave breaking. When breaking appears two stationary points satisfy equation (9.96). Those points are x_1 and x_2.

In this case the solution to the Burgers equation (9.18) takes the form

$$u(x,t) = \frac{\displaystyle\sum_{i=1,2} \frac{x-x_i}{t} \frac{\exp\left(-\frac{(x-x_i)^2}{4t} - \frac{1}{2}\int_{-\infty}^{x_i} u_0(s)ds\right)}{(u_0''(x_i))^{1/2}}}{\displaystyle\sum_{i=1,2} \frac{\exp\left(-\frac{(x-x_i)^2}{4t} - \frac{1}{2}\int_{-\infty}^{x_i} u_0(s)ds\right)}{(u_0''(x_i))^{1/2}}}. \tag{9.100}$$

Far from the region of the wavebreaking (that is far from the region of the front of the shock wave) in the nominator and denominator of expression (9.100) one of the exponent is dominant. In that region, where we suppose that the stationary point is x_1, we have

$$u_0(x_1) \approx \frac{x-x_1}{t}, \qquad \text{and} \qquad x = x_1 + u_0(x_1)t, \tag{9.101}$$

while in the region where the stationary point is x_2 we have

$$u_0(x_2) \approx \frac{x-x_2}{t}, \qquad \text{and} \qquad x = x_2 + u_0(x_2)t. \tag{9.102}$$

Transition from the first solution to the second one takes place where both exponents are equal each other. In this point we have

$$\frac{(x-x_1)^2}{2t} + \int_{-\infty}^{x_1} u_0(s)ds = \frac{(x-x_2)^2}{2t} + \int_{-\infty}^{x_2} u_0(s)ds. \tag{9.103}$$

This expression can be rewritten as

$$\frac{(x-x_1)^2 - (x-x_2)^2}{2t} = \int_{x_1}^{x_2} u_0(s)ds. \tag{9.104}$$

Since $(x - x_i)/t = u_0(x_i)$ we rewrite expression (9.104) as

$$\frac{(u_0(x_1) - u_0(x_2))(x_1 - x_2)}{2t} = \int_{x_2}^{x_1} u_0(s)ds. \qquad (9.105)$$

This relationship corresponds to the rule of surface area conservation as it is illustrated in Fig.9.4.

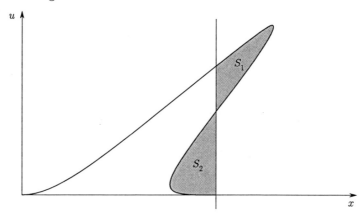

Fig. 9.4. The rule of the surface area conservation.

The surfaces of shaded areas are equal. The surface area conservation rule gives us a receipe of how to find the position of the shock wave front after the wave breaking.

Exercise. Demonstrate that equation (9.95) can also be represented in the form

$$u(x,t) = \frac{\int_{-\infty}^{+\infty} u_0(x') \exp\left(-\frac{(x - x')^2}{4t} - \frac{1}{2}\int_{-\infty}^{x'} u_0(x'')dx''\right) dx'}{\int_{-\infty}^{+\infty} \exp\left(-\frac{(x - x')^2}{4t} - \frac{1}{2}\int_{-\infty}^{x'} u_0(x'')dx''\right) dx'}. \qquad (9.106)$$

◇

Example. We consider the Cauchy problem for which

$$u(x,0) = \delta(x), \qquad (9.107)$$

where $\delta(x)$ is the Dirac delta function. After the integration of equation (9.106) we obtain the solution to the Burgers equation (9.18) with the initial condition (9.107) in the form

$$u(x,t) = \frac{2}{\left(\frac{1}{(\sqrt{e} - 1)} + \frac{1}{2}\mathrm{erfc}\left(\frac{x}{\sqrt{4t}}\right)\right)} \left(\frac{1}{\sqrt{4\pi t}}\exp\left(\frac{x^2}{4t}\right)\right). \qquad (9.108)$$

Here erfc(x) is the complementary error function:

$$\text{erfc(x)} = \frac{2}{\pi^{1/2}} \int\limits_{x}^{\infty} \exp(-s^2)ds. \tag{9.109}$$

The solution given by expression (9.108) comprises he product of two factors: the fundamental solution to diffusion equation and the predexponent which represents the effects of nonlinearity in the Burgers equation.

For small values of time t, the variations of the predexponent factor are small. The expression (9.108) resembles the fundamental solution to the diffusion equation. With increasing time the effect of nonlinearity becomes stronger and the profile $u(x)$ becomes nonsymmetric.

•

Exercise. Consider the solution to the Burgers equation which describes the N-wave, to which in variable $z(x,t)$ corresponds the supplementary solution to the diffusion equation:

$$z(x,t) = \frac{x}{t} \frac{1}{(2\pi t)^{1/2}} \exp\left(\frac{x^2}{4t}\right). \tag{9.110}$$

◇

Exercise. Consider solution to the Burgers equation which describes the interacting shock waves.

◇

Corollary. We can observe that the Cole-Hopf transformation of the propagating wave $u(x - Vt)$ (see equation (9.56)) results in the problem to find the function $z(x,t)$. For the boundary conditions (9.52) we have

$$z(x,t) = \exp\left(-\frac{u_1}{2}(x - u_1 t)\right) + \exp\left(-\frac{u_2}{2}(x - u_2 t)\right). \tag{9.111}$$

Thus, the stationary solution to the Burgers equation corresponds to the linear combination of the exponents.

⋆

Chapter 10

Nonlinear Waves in Dispersive Media

Nonlinear waves in dispersive media. Canonical equations describing nonlinear waves. Korteweg-de Vries equation. Stationary solution to the Korteweg-de Vries equation. Solitons. Collisionless shock waves. Multisoliton solution of the KdV equation. The Hirota Transformation. The nonlinear Schrödinger equation. Wave behavior near the focus. The Lighthil criterion of the modulation instability. Aberrationless regime of the self-focusing of electromagnetic radiation. Solitons described by nonlinear Schrödinger equation.

***

10.1   Canonical Equations Describing Nonlinear Waves

The standard set of the equations that describe the nonlinear waves in dispersive media starts from:

1. The Korteweg-de Vries equation:

$$\partial_t u + u \partial_x u + \beta \partial_{xxx} u = 0; \tag{10.1}$$

2. The nonlinear Schrödinger equation:

$$i \partial_t u + \partial_{xx} u + \nu |u|^2 u = 0; \tag{10.2}$$

3. The sine-Gordon equation:

$$\partial_{tt} u - \partial_{xx} u + \sin u = 0. \tag{10.3}$$

The set also contains many other nonlinear equations in partial derivatives that are of generic interest.

10.2   The Korteweg-de Vries Equation

The Korteweg-de Vries equation (10.1) describes simultaneous action of the hydrodynamic nonlinearity, $u\partial_x u$, and dispersion, $\beta\,\partial_{xxx} u$. This equation can be derived in the frame of the "shallow water" approximation. The finite depth of the water being taken into account leads to appearance of high order derivatives in the equation, which describe the wave dispersion. Then, we need to perform the expansion of the series to the value of the perturbation amplitude and to use the multiple scale expansion method similar to the one we used to obtain the Burgers equation. As a result we obtain an equation of the form

$$\partial_t \eta + c_0\left(\partial_x \eta + \frac{3}{2h_0}\eta\partial_x\eta + \frac{h_0^2}{6}\partial_{xxx}\eta\right) = 0. \tag{10.4}$$

Here x is the coordinate along the water layer, t is time, η is the displacement of the water surface from its equilibrium position, h_0 is the water depth, $c_0 = (gh_0)^{1/2}$ is the velocity of propagation of linear waves (see equation (7.22)). We introduce a new dependent variable, $\eta = Au + B$, and new independent variables x/x_0 and t/t_0 and use notations x and t for them. We choose the constants A, B, x_0 and t_0 as it follows.

$$t_0 = h_0/c_0, \quad x_0 = h_0/6^{1/3}, \quad B = -2h_0/3, \quad A = -4h_0/6^{1/3}. \tag{10.5}$$

As a result equation (10.4) takes the form

$$\partial_t u - 6u\partial_x u + \partial_{xxx} u = 0, \tag{10.6}$$

which sometimes is more convenient than (10.1) and (10.4).

10.2.1   The Stationary Solution to the Korteweg-de Vries Equation

We look for the solution to equation (10.1) in the form of the uniform propagating wave. The velocity of propagation is V, that is, the solution is $u(x - Vt)$. From equation (10.1) it follows that

$$-Vu' - uu' + \beta u''' = 0, \tag{10.7}$$

where the prime denotes the derivative with respect to the coordinate

$$X = x - Vt. \tag{10.8}$$

We note that equation (10.7) is invariant with respect to the transformation

$$u \to u + v, \qquad V \to V + v. \tag{10.9}$$

Integrating the ones in equation (10.7) we find

$$\beta u'' - Vu + \frac{u^2}{2} = \text{const.} \tag{10.10}$$

This equation is identical to the equation that describes the nonlinear oscillator and has been discussed above. Equation (10.10) has the energy integral which we can write in the form

$$(u')^2 = -\frac{1}{3\beta}(u - u_1)(u - u_2)(u - u_3),\qquad(10.11)$$

where the constant values u_1, u_2, and u_3 are determined by intial conditions and related to the velocity of the wave propagation as

$$V = \frac{1}{3}(u_1 + u_2 + u_3).\qquad(10.12)$$

The effective potential energy of the oscillator is equal to $(u-u_1)(u-u_2)(u-u_3)/3\beta$, and is shown in Fig. 10.1. as a function of u

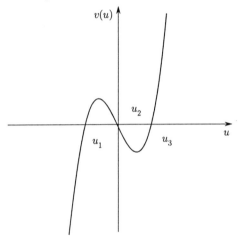

Fig. 10.1. Effective potential in equation (10.11).

Using the property of the Korteweg-de Vries equation (10.1) to be invariant under transformation (10.9), we choose the constant u_3 to be equal to zero, $u_3 = 0$. As a result we obtain the effective potential energy equal to

$$V_{\text{eff}} = \frac{1}{3\beta}u(u_1 - u)(u - u_2).\qquad(10.13)$$

The phase plane of equation (10.11) that corresponds to this potential is shown in Fig. 10.2.

We see that the system has two critical points where $dV_{\text{eff}}/du = 0$. The first is of the elliptic type, while the second one is the saddle point. There is a separatrix which corresponds to the saddle point.

The solution to equation (10.11) with $u_3 = 0$ has the form

$$X = x - Vt = \int_u^{u_1} \frac{(3\beta)^{1/2}ds}{(s(u_1 - s)(s - u_2))^{1/2}} = \left(\frac{12\beta}{u_1}\right)^{1/2} F(\varphi, \kappa).\qquad(10.14)$$

u'

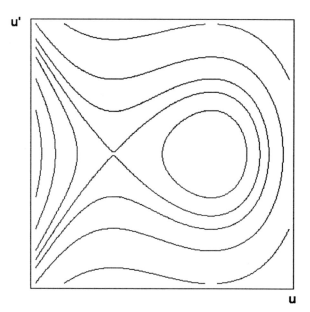

u

Fig. 10.2. Phase plane of equation (10.11).

Here $F(\kappa, \varphi)$ is the elliptic integral of the first kind:

$$F(\varphi, \kappa) = \int_0^{\varphi} \frac{ds}{(1 - \kappa^2 \sin^2 s)^{1/2}}, \tag{10.15}$$

where

$$\sin \varphi = \left(\frac{u_1 - u}{u_1 - u_2} \right)^{1/2}, \quad \text{and} \quad \kappa = \left(1 - \frac{u_2}{u_1} \right)^{1/2}. \tag{10.16}$$

From expression (10.14) it follows that

$$u(x, t) = u(x - Vt) = u_1 \mathrm{dn}^2 \left(\left(\frac{u_1}{12\beta} \right)^{1/2} X, \kappa \right). \tag{10.17}$$

Here $\mathrm{dn}(y, \kappa)$ is the Jacobi elliptic function. The Jacobi elliptic function is periodic in the variable y, that is, the solution given by relationship (10.17) is periodic in the variable X.

Expression (10.17) describes a nonlinear wave with the wavelength, which corresponds to the period of nonlinear oscillations discussed above. The wavelength value is equal to

$$\lambda = 4 \left(\frac{3\beta}{u_1} \right)^{1/2} F \left(\frac{\pi}{2}, \kappa \right) \equiv 4 \left(\frac{3\beta}{u_1} \right)^{1/2} K(\kappa), \tag{10.18}$$

with $K(\kappa)$ being equal to the complete elliptic integral of the first kind,

$$K(\kappa) = \int\limits_{0}^{\pi/2} \frac{ds}{(1 - \kappa^2 \sin^2 s)^{1/2}}. \tag{10.19}$$

The amplitude of the wave as it follows from equations (10.13) and (10.17) is equal to $(u_1 - u_2)$. In the limit of small values of the wave amplitude, $|u_1 - u_2| \ll 1$, expression (10.17) gives

$$u(X) \approx \frac{u_1 + u_2}{2} + \frac{u_1 - u_2}{2} \cos kX, \tag{10.20}$$

where the wavenumber is

$$k = \left(\frac{u_1}{3\beta}\right)^{1/2}. \tag{10.21}$$

When the constant u_2 tends to zero, the value of κ tends to one. For $\kappa \to 1$ the complete elliptic integral of the first kind, $K(\kappa)$, approximately equals to

$$K(\kappa) \approx \frac{1}{2} \ln \frac{16}{1 - \kappa^2}. \tag{10.22}$$

This limit corresponds to the motion in the vicinity of the separatrix trajectory in the phase plane (see Fig. 10.2).

To analyse the behavior of the solution on the separatrix we assume u_2 to be equal to zero. In this case, from equations (10.11) and (10.13) we have

$$(u')^2 = \frac{u^2(3V - u)}{6\beta}. \tag{10.23}$$

We suppose that both V and β are positive, $V > 0$ and $\beta > 0$. That corresponds to the nonlinear wave with a positive amplitude, which propagates in the positive direction along the x-coordinate. The maximum value of the wave amplitude, $u = u_{\max}$, as it follows from equation (10.23), is

$$u_{\max} = 3V \equiv u_1. \tag{10.24}$$

Since in the course of the motion along the separatrix it takes infinite time for the solution to get into saddle point, $u = 0$, and respectively to leave saddle point, the solution to equation (10.23),

$$u(x - Vt) = \frac{3V}{\cosh^2\left(\frac{(x - Vt)}{2}\left(\frac{V}{\beta}\right)^{1/2}\right)}, \tag{10.25}$$

describes the solitary wave, which is also called the soliton. Its amplitude and the propagation velocity are related each other, as well as its width, $k^{-1} = 2(\beta/V)^{1/2}$, depends on the soliton amplitude, as it is seen from formula (10.25).

10.2.2   Collisionless Shock Waves

Here we discuss the results of joint action of the dispersion and dissipation on the structure of nonlinear waves. These effects are described in the frame of the Korteweg-de Veries-Burgers equation:

$$\partial_t u + u \partial_x u - \nu \partial_{xx} u + \beta \partial_{xxx} u = 0. \tag{10.26}$$

We look for a solution to this equation in the form of stationary wave propagating with constant velocity. The solution depends on the variable $X = x - Vt$. In this case u(X) obeys an ordinary differential equation

$$u''' - \frac{\nu}{\beta} u'' + \frac{1}{\beta} u u' - \frac{V}{\beta} u' = 0. \tag{10.27}$$

Integrating this equation ones we obtain

$$u'' - \frac{\nu}{\beta} u' + \frac{1}{2\beta} u^2 - \frac{V}{\beta} u = \text{const.} \tag{10.28}$$

This is the equation for a nonlinear oscillator with friction. Above we have already discussed behavior of solutions to similar equations. The behavior of the solution is illustrated in Fig.10.3.

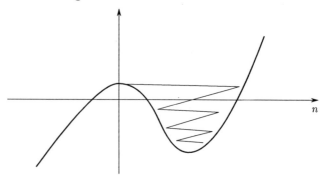

Fig. 10.3. Damping oscillations described by equation (10.28).

These solutions describe the change of the amplitude of the wave from zero far ahead of the shock wave front, to $u_1 = 2V$ far behind the shock wave front.

The decay of oscillation amplitude, with the coefficient equal ν/β , results in the decreasing of the amplitude of solitons as it is shown in Fig.10.4 a),b).

If dissipation effects are more important than the effects of dispersion, $\nu/\beta \gg 1$, there are no oscillations at the shock wave front. More precisely, the decay should be large enough, $\nu \gg \nu_{cr}$, with

$$\nu_{cr} = (4\beta u_1)^{1/2}. \tag{10.29}$$

In this case the wave has a monotonous structure.

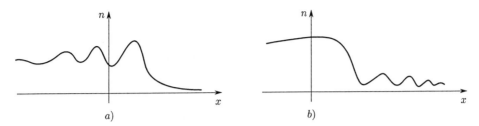

Fig. 10.4. Structure of the shock wave front.

In the case when $\nu/\beta \ll 1$, the dispersion effects are dominant and there are many well seen solitons near the front. For $\beta > 0$ the oscillations are localized behind the front (Fig. 10.4 a), while for $\beta < 0$ they are ahead the front (Fig.10.4 b).

Example. We consider the magnetoacoustic waves in a plasma. If the wave propagates almost perpendicularly to the direction of the magnetic field, its frequency and wavenumber obey the dispersion equation

$$\omega^2(k) \simeq k^2 v_a^2 \left(1 - \frac{k^2 c^2}{\omega_{pe}^2}\right), \tag{10.30}$$

where the Alfvèn velocity, v_a, and the Langmuir frequency, ω_{pe}, are equal to $v_a = B/(4\pi\rho)^{1/2}$, and $\omega_{pe} = (4\pi n e^2/m)^{1/2}$, respectively. It is supposed that $kc \ll \omega_{pe}$. From these expressions it follows that $\omega \approx k v_a - k^3 v_a c^2/2\omega_{pe}^2$, that is

$$\beta \approx \frac{v_a c^2}{2\omega_{pe}^2} > 0. \tag{10.31}$$

This means that the oscillations are localized behind the front of the magnetoacoustic shock wave propagating perpendicularly to the magnetic field. When the direction of the magnetoacoustic wave propagation is almost parallel to the direction of the magnetic field, the wave frequency and wavenumber are related via the dispersion equation

$$\omega^2(k) \simeq k^2 v_a^2 \left(1 + \frac{k^2 c^2 \cos^2\theta}{\omega_{pi}^2}\right), \tag{10.32}$$

where the frequency, ω_{pi}, is equal to $\omega_{pi} = (4\pi n e^2/M)^{1/2}$, with M being the ion mass. In this case it is supposed that $kc \ll \omega_{pi}$. From these expressions it follows that $\omega \approx k v_a + k^3 v_a c^2/2\omega_{pi}^2$, that is

$$\beta \approx -\frac{v_a c^2}{2\omega_{pi}} < 0. \tag{10.33}$$

This means the oscillations at the front of the magnetoacoustic shock wave, propagating in the quasi-parallel with respect to the magnetic field, are localized ahead the front.

10.2.3 Multisoliton Solution of the KdV Equation. The Hirota Transformation.

Below we shall discuss how to find a solution to Caushy problem for the Korteweg-de Vries equation with the inverse scattering method. In a particular case when the solution comprises a finite number of solitary waves, it can be obtained with the Hirota transformation.

As the first step, we change the dependent variable u in the Korteweg-de Vries equation according to

$$u = \partial_x p. \tag{10.34}$$

Then, integrating the Korteweg-de Vries equation (10.1) with respect to x, we find

$$\partial_t p + \frac{1}{2} (\partial_x p)^2 + \partial_{xxx} p = 0. \tag{10.35}$$

Here and below we assume $\beta = 1$.

Then, as the second step, we introduce the function $F(x,t)$, which in terms of $p(x,t)$ is expressed as

$$p = 12 \partial_x (\ln F). \tag{10.36}$$

From equations (10.35) and (10.36) we obtain

$$F \partial_x (\partial_t F + \partial_{xxx} F) - \partial_x F (\partial_t F + \partial_{xxx} F) +$$

$$+ 3 \left((\partial_{xx} F)^2 - \partial_x F \partial_{xxxx} F \right) = 0. \tag{10.37}$$

We see that the equation contains the operator

$$\partial_t + \partial_{xxx} \tag{10.38}$$

of the Korteweg-de Vries equation.

Now we rewrite the one-soliton solution given by expression

$$u(x - Vt) = \frac{3V}{\cosh^2 \left(\dfrac{(x - Vt)}{2} \left(\dfrac{V}{\beta} \right)^{1/2} \right)}, \tag{10.39}$$

in the form

$$u = \frac{3\alpha^2}{\cosh^2 \left(\dfrac{\theta - \theta_0}{2} \right)}, \tag{10.40}$$

where the phase θ is

$$\theta = \alpha x - \alpha^3 t, \tag{10.41}$$

and α and θ_0 are the parameters of the solution.

This one-soliton solution can also be represented in the form

$$u = \partial_x \left(6\alpha \tanh \left(\frac{\theta - \theta_0}{2} \right) - 1 \right) = 12 \partial_{xx} (\ln F) \tag{10.42}$$

with
$$F = 1 + \exp(-(\theta - \theta_0)) = 1 + \exp(-\alpha(x - s) + \alpha^3 t). \tag{10.43}$$

Here $s = \theta_0/\alpha$.

As we see, the function (10.43) obeys the equation

$$\partial_t F + \partial_{xxx} F = 0. \tag{10.44}$$

In a linear approximation with respect to the wave amplitude, if it is valid, one could seek the solution in the form of the expansion

$$F = 1 + F^{(1)} + F^{(2)} + \dots . \tag{10.45}$$

Substituting the expansion into equation (10.37) we see that functions $F^{(i)}$ satisfy an infinite system of equations:

$$\partial_t F^{(1)} + \partial_{xxx} F^{(1)} = 0,$$

$$F^{(0)} \partial_x \left(\partial_t F^{(2)} + \partial_{xxx} F^{(2)} \right) = -3 \left(\left(\partial_{xx} F^{(1)} \right)^2 - \partial_x F^{(1)} \partial_{xxxx} F^{(1)} \right),$$

$$-\;-. \tag{10.46}$$

If we choose $F^{(1)}$ being equal

$$F^{(1)} = f_1 + f_2, \tag{10.47}$$

where

$$f_j = \exp(-\alpha_j(x - s_j) + \alpha_j^3 t), \qquad j = 1, 2, \tag{10.48}$$

then for $F^{(2)}$ we obtain the equation

$$\partial_x \left(\partial_t F^{(2)} + \partial_{xxx} F^{(2)} \right) = 3\alpha_1 \alpha_2 (\alpha_1 - \alpha_2)^2 f_1 f_2. \tag{10.49}$$

The remaining equations in the system (10.46) vanish. As a result, from equation (10.49) we obtain

$$F = 1 + f_1 + f_2 + \left(\frac{\alpha_1 - \alpha_2}{\alpha_1 + \alpha_2} \right)^2 f_1 f_2. \tag{10.50}$$

This expression leads to the conclusion that the N−exponent (with $f_1, f_2, f_3, \dots f_N$) function F can be written as

$$F = \det(F_{nm}), \tag{10.51}$$

where the matrix F_{nm} equals

$$F_{nm} = \delta_{nm} + \frac{2\alpha_m}{\alpha_m + \alpha_n} f_m. \tag{10.52}$$

As an example, let us consider a two-soliton solution (10.50). Using expressions (10.34) and (10.36) we find

$$\frac{u}{12} = \frac{\alpha_1^2 f_1 + \alpha_2^2 f_2 + 2(\alpha_1 - \alpha_2)^2 f_1 f_2 - \dfrac{\alpha_1 - \alpha_2}{(\alpha_1 + \alpha_2)^2}(\alpha_2^2 f_1^2 f_2 + \alpha_1^2 f_1 f_2^2)}{\left(1 + f_1 + f_2 + \left(\dfrac{\alpha_1 - \alpha_2}{\alpha_1 + \alpha_2}\right)^2 f_1 f_2\right)^2},$$

(10.53)

where f_j are given by expressions (10.48).

The one-soliton solution is described by the formula

$$u = \frac{12\alpha^2 f}{(1 + f)^2};$$

(10.54)

here the amplitude maximum value is equal to $u_m = 3\alpha^2$; the soliton is localized at $x - \alpha^3 t = s$; the velocity of the soliton propagation is $V = \alpha^3$.

Let us consider expression (10.53) in the limits when one of the exponents is dominant, or is much less than one.

1. When $f_1 \approx 1$ and $f_2 \ll 1$ we have

$$u \approx \frac{12\alpha_1^2 f_1}{(1 + f_1)^2};$$

(10.55)

this expression describes the soliton at the coordinate $x - \alpha_1^3 t = s_1$.

2. When $f_1 \approx 1$ and $f_2 \gg 1$ we have

$$u \approx \frac{12\alpha_1^2 \tilde{f}_1}{(1 + \tilde{f}_1)^2}$$

(10.56)

with

$$\tilde{f}_1 = \left(\frac{\alpha_1 - \alpha_2}{\alpha_1 + \alpha_2}\right)^2 f_1.$$

(10.57)

This formula corresponds to the soliton at the coordinate $x - \alpha_1^3 t = \tilde{s}_1$, where

$$\tilde{s}_1 = s_1 - \frac{2}{\alpha_1} \ln \left|\frac{\alpha_1 + \alpha_2}{\alpha_1 - \alpha_2}\right|.$$

(10.58)

From equation (10.53) it follows that the interaction between two solitons leads to the displacement of the solitons with respect to their initial position as it is shown in Fig. 10.5 a), b).

Let us consider the asymptotics of relationship (10.53) for $t \to -\infty$. The first soliton has parameters:

$$\alpha_1, \qquad x = s_1 + \alpha_1^3 t, \qquad f_1 \approx 1, \qquad f_2 \ll 1. \tag{10.59}$$

The second soliton has parameters:

$$\alpha_2, \qquad x = s_2 + \frac{2}{\alpha_2} \ln \left|\frac{\alpha_1 + \alpha_2}{\alpha_1 - \alpha_2}\right| + \alpha_2^3 t, \qquad f_1 \gg 1, \qquad f_2 \approx 1. \tag{10.60}$$

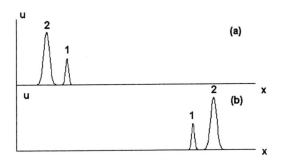

Fig. 10.5. Interacting solitons.

When $t \to +\infty$, for the first soliton we have

$$\alpha_1, \qquad x = s_1 - \frac{2}{\alpha_1} \ln \left| \frac{\alpha_1 + \alpha_2}{\alpha_1 - \alpha_2} \right| + \alpha_1^3 t, \qquad f_1 \approx 1, \qquad f_2 \gg 1, \qquad (10.61)$$

and for the second one we have

$$\alpha_2, \qquad x = s_2 + \alpha_2^3 t, \qquad f_1 \ll 1, \qquad f_2 \approx 1. \qquad (10.62)$$

We see that the interaction between solitons results in the displacement of the first soliton to

$$-\frac{2}{\alpha_1} \ln \left| \frac{\alpha_1 + \alpha_2}{\alpha_1 - \alpha_2} \right| \qquad (10.63)$$

and of the second soliton to

$$\frac{2}{\alpha_2} \ln \left| \frac{\alpha_1 + \alpha_2}{\alpha_1 - \alpha_2} \right|. \qquad (10.64)$$

The interaction ("collision") takes place at time

$$t = \frac{s_1 - s_2}{\alpha_2^2 - \alpha_1^2}, \qquad (10.65)$$

and at the x-coordinate equal

$$x = \frac{\alpha_2^2 s_1 - \alpha_1^2 s_2}{\alpha_2^2 - \alpha_1^2}, \qquad (10.66)$$

where $f_1 \approx 1$ and $f_2 \approx 1$.

Exercise. Analyse the cases when $\alpha_1 \approx \alpha_2$ and $\alpha_1 \gg \alpha_2$.

◇

10.3 The Nonlinear Schrödinger Equation

The nonlinear Schrödinger equation describes the self-focusing of electromagnetic waves and self-modulation of wave packets. This equation provides an example of so-called parabolic equations to describe wave packet propagation, when one considers a wave centred around the frequency ω_0 and wave number k_0. In the linear approximation the parabolic equation describes the wave near the focus.

10.3.1 Wave Behavior near the Focus

We consider the wave behavior near the focus and seek the solution for the problem of the wave propagation in a quasistationary, nonuniform regime. The wave focusing is described by the wave equation

$$\partial_{tt}u - c^2\Delta u = 0. \tag{10.67}$$

Assuming that time dependence of the wave is given by $u = u\exp(-i\omega t)$ from equation (10.67) we obtain

$$\Delta u + k^2 u = 0, \tag{10.68}$$

where $k^2 = \omega^2/c^2$. Then we consider the case when the solution of equation (10.68) can be represented in the form

$$u(\mathbf{x}) = U(x, y, z)\exp(ik\mathbf{x}), \tag{10.69}$$

assuming that the scale of the amplitude, U, nonhomogeneity is much larger then $2\pi/k$, that is $\partial^2/\partial x^2 \ll k^2$. In this approximation we obtain

$$2ik\partial_x U + \Delta_\perp U = 0. \tag{10.70}$$

This is formally the Schrödinger equation. The solution to this equation is

$$U(x, r_\perp) = U_0 \frac{kl^2}{\pi(x - ix_0)}\exp\left(-\frac{ikr_\perp^2}{2(x - ix_0)}\right). \tag{10.71}$$

We see that the wave reaches the maximum $U_{max} \approx U_0 kl^2/\pi x_0$ at $x = 0, r_\perp = 0$. The typical size of the wave beam in the transverse direction is $\delta r_\perp \approx (x_0/k)^{1/2}$. This relationship can be rewritten for the size of the region in the longitudinal direction

$$x_0 \approx \frac{2\pi\delta r_\perp^2}{\lambda} \equiv l_R, \tag{10.72}$$

with l_R being the Rayleigh length, and $U_{max} \approx U_0(l/\delta r_\perp)^2$.

10.3.2 Nonlinear Schrödinger Equation Obtained with Multiple Scale Expansions

We consider a nonlinear wave equation

$$\partial_{tt}u - \partial_{xx}u = F(u), \tag{10.73}$$

where, assuming the wave amplitude to be small $u = O(\varepsilon)$, the function $F(u)$ can be expanded as

$$F(u) = -u + \alpha u^3 + \dots. \tag{10.74}$$

It means that we assume the function $F(u)$ to be an add function of the argument u. We look for a solution of equation (10.73) in the form

$$u = \varepsilon u^{(1)} + \varepsilon^2 u^{(2)} + \varepsilon^3 u^{(3)} + \dots, \tag{10.75}$$

expanding time and space derivatives as

$$\partial_t = \partial_{t_0} + \varepsilon \partial_{t_1} + \varepsilon^2 \partial_{t_2} + \varepsilon^3 \partial_{t_3} + \dots, \tag{10.76}$$

and

$$\partial_x = \partial_{x_0} + \varepsilon \partial_{x_1} + \varepsilon^2 \partial_{x_2} + \varepsilon^3 \partial_{x_3} + \dots, \tag{10.77}$$

respectively. From this we have

$$\partial_{tt} = \partial_{t_0 t_0} + 2\varepsilon \partial_{t_0 t_1} + 2\varepsilon^2 \partial_{t_0 t_2} + \varepsilon^2 \partial_{t_1 t_1} + \dots, \tag{10.78}$$

$$\partial_{xx} = \partial_{x_0 x_0} + 2\varepsilon \partial_{x_0 x_1} + 2\varepsilon^2 \partial_{x_0 x_2} + \varepsilon^2 \partial_{x_1 x_1} + \dots, \tag{10.79}$$

$$F(u) = \varepsilon u^{(1)} + \varepsilon^2 u^{(2)} + \varepsilon^3 u^{(3)} + \varepsilon^3 \alpha u^{(1)3} + \dots. \tag{10.80}$$

Substituting these relationships into equation (10.73) we obtain to the first order in ε:

$$\partial_{t_0 t_0} u^{(1)} - \partial_{x_0 x_0} u^{(1)} + u^{(1)} = 0. \tag{10.81}$$

We write the solution to this order in the form

$$u^{(1)} = U(x_1, x_2, \dots, t_1, t_2, \dots) \exp(i\theta) + c.c., \tag{10.82}$$

where "c.c." stands for complex conjugate, $\theta = kx_0 - \omega t_0$ and

$$\omega^2 = k^2 + 1. \tag{10.83}$$

To the second order in ε we have

$$\partial_{t_0 t_0} u^{(2)} - \partial_{x_0 x_0} u^{(2)} + u^{(2)}$$

$$= -2\partial_{t_0 t_1} U(x_1, \dots, t_1, \dots) \exp(i\theta) + 2\partial_{x_0 x_1} U(x_1, \dots, t_1, \dots) \exp(i\theta). \tag{10.84}$$

Eliminating the secular terms in the right-hand side of the equation we obtain a relationship

$$k\partial_{x_1} U - \omega \partial_{t_1} U = 0. \tag{10.85}$$

Nonlinear terms come to play to the third order in ε, when we have

$$\partial_{t_0 t_0} u^{(3)} - \partial_{x_0 x_0} u^{(3)} + u^{(3)}$$

$$= -2\partial_{t_0 t_2} U \exp(i\theta) + 2\partial_{x_0 x_2} U \exp(i\theta)$$

$$- \partial_{t_1 t_1} U \exp(i\theta) + \partial_{x_1 x_1} U \exp(i\theta) + \alpha (U \exp(i\theta) + c.c.)^3. \tag{10.86}$$

The last term in the right-hand side is equal to

$$(U \exp(i\theta) + U^* \exp(-i\theta))^3$$

$$= U^3 \exp(3i\theta) + 3U^2 U^* \exp(i\theta) + 3U(U^*)^2 \exp(-i\theta) + (U^*)^3 \exp(-3i\theta). \quad (10.87)$$

Substituting this expression into equation (10.86) and eliminating the secular terms we obtain

$$2i\omega\partial_{t_2}U + 2ik\partial_{x_2}U - \partial_{t_1 t_1}U + \partial_{x_1 x_1}U + 3\alpha U^2 U^* = 0. \quad (10.88)$$

Using relationships $\partial_{t_1}U = (k/\omega)\partial_{x_1}U$, $\partial_{t_1 t_1}U = (k/\omega)^2\partial_{x_1 x_1}U$, $\partial_{t_1} = \varepsilon\partial_t$, $\partial_{x_1} = \varepsilon\partial_x$, $\partial_{t_2} = \varepsilon^2\partial_t$, $\partial_{x_2} = \varepsilon^2\partial_x$, and the dispertion equation (10.83), we find

$$2i\omega(\partial_t + \frac{k}{\omega}\partial_x)U + \frac{1}{\omega^2}\partial_{xx}U + 3\alpha|U|^2U = 0. \quad (10.89)$$

Normalizing variables, in the frame moving with the group velocity $v_g = k/\omega$ we rewrite the nonlinear Schrödinger equation as

$$i\partial_t u + \partial_{xx}u + \nu|u|^2 u = 0. \quad (10.90)$$

10.3.3   The Lighthil Criterion of the Modulation Instability

The dispersion equation that corresponds to equation (10.90) can be written as

$$\omega(k) = k^2 - \nu|u|^2. \quad (10.91)$$

The nonlinear part of equation (10.91) can be regarded as a nonlinear shift of the frequency which is proportional to the square of the wave amplitude. (Compare with the nonlinear shift of the frequency of nonlinear oscillator given by relationship $\omega = \omega_0 + \varepsilon a^2$).

Restoring dimensional units, the dispersion equation (10.91) in the laboratory frame can be written as

$$\omega(k) = W'k + \frac{1}{2}W''k^2 - \nu|u|^2. \quad (10.92)$$

The corresponding form of a nonlinear Schrödinger equation is

$$i\left(\partial_t u + W'\partial_x u\right) + \frac{1}{2}W''\partial_{xx}u + \nu|u|^2 u = 0. \quad (10.93)$$

Linearizing this equation around a uniform equilibrium, $u = u_0(t)$, we can demonstrate that for

$$\nu W'' > 0 \quad (10.94)$$

we have instability, while for

$$\nu W'' < 0 \quad (10.95)$$

the equilibrium is stable. These conditions are known as the Lighthill criterion of modulation instability.

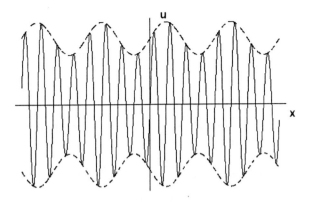

Fig. 10.6. Modulation instability.

We may also obtain this criterion by considering the equation for the wave energy propagation

$$\partial_t a^2 + \partial_x(v_g a^2) = 0, \tag{10.96}$$

with

$$\partial_t k + \partial_x \omega = 0, \tag{10.97}$$

where we take into account the nonlinear frequency shift: $\omega = W + \nu a^2$. We rewrite the system of equations (10.96), (10.97) as

$$\partial_t k + W' \partial_x k + \nu \partial_x a^2 = 0, \tag{10.98}$$

$$\partial_t a^2 + W' \partial_x a^2 + W'' a^2 \partial_x k = 0. \tag{10.99}$$

The equation of the characteristics of this system has the form

$$\frac{dx}{dt} = W' \pm (-\nu W'')^{1/2} a. \tag{10.100}$$

We see that the characteristics are real if $\nu W'' < 0$, that is, the wave is stable against the modulational instability if $\nu W'' < 0$ and unstable for $\nu W'' > 0$.

10.3.4 Aberrationless Regime of the Self-Focusing of Electromagnetic Radiation

The self-focusing of electromagnetic radiation can be described analytically in some regimes with the nonlinear Schrödinger equation which we write in the 3D case as

$$2i\partial_t u + \Delta_\perp u + |u|^2 u = 0. \tag{10.101}$$

The last term in the left-hand side of Eq. (10.101) corresponds to the nonlinear shift of the electromagnetic radiation frequency in a plasma, which is proportional to the second power of the wave amplitude, $|u|^2$. Above we have already discussed the role of the remained terms in this equation that correspond to the linear part of the equation .

We represent the nonlinear Schrödinger equation in the so-called "hydrodynamics form". We write the function $u(\mathbf{r}_\perp, t)$ as

$$u(\mathbf{r}_\perp, t) = a(\mathbf{r}_\perp, t) \exp\left(i\theta(\mathbf{r}_\perp, t)\right), \tag{10.102}$$

where $a(\mathbf{r}_\perp, t)$ and $\theta(\mathbf{r}_\perp, t)$ are real. Substituting this representation of function $u(\mathbf{r}_\perp, t)$ into equation (10.101) we obtain

$$2\partial_t a + 2\nabla_\perp a \cdot \nabla_\perp \theta + a\Delta_\perp \theta = 0, \tag{10.103}$$

$$2\partial_t \theta + (\nabla_\perp \theta)^2 = a^2 + \frac{1}{a^2}\Delta_\perp a. \tag{10.104}$$

Here

$$\nabla_\perp = \mathbf{e}_y \partial_y + \mathbf{e}_z \partial_z, \qquad \Delta_\perp = \partial_{yy} + \partial_{zz}.$$

In the aberrationless (longwavelength) limit, we can neglect the last term in the right hand-side of equation (10.104). As a result equations (10.103) and (10.104) take the form of equations of gas dynamics

$$\partial_t \nu + \mathrm{div}_\perp(\nu \mathbf{u}) = 0, \tag{10.105}$$

$$\partial_t \mathbf{u} + (\mathbf{u}\nabla_\perp)\mathbf{u} = \nabla_\perp \nu, \tag{10.106}$$

where $\nu = a^2$ and $\mathbf{u} = \nabla_\perp \theta$. We see that the term in the right hand-side of the second equation corresponds to the instability with the growth rate $\gamma = k_\perp \nu_0$. This describes in the longwavelength limit the wave self-focusing (*Askar'yan, 1962*).

10.3.5 Solitons Described by Nonlinear Schrödinger Equation

We look for a solution to equation (10.90) in the form of a uniform propagating with a constant velocity wave of the form

$$u(x, t) = a(X) \exp(i(\delta k x - \delta \omega t)), \tag{10.107}$$

where $X = x - Vt$.

Substituting this expression into equation (10.90), we obtain

$$a'' - i\left(\delta k - \frac{V}{2}\right)a' + (\delta\omega - (\delta k)^2)a + \nu|a|^2 a = 0. \tag{10.108}$$

Now we choose the values of δk and $\delta \omega$ to be equal to

$$\delta k = \frac{V}{2}, \qquad \delta\omega = \frac{V^2}{4} - \alpha, \tag{10.109}$$

with α being a constant.

As a result, the equation (10.108) transforms into

$$a'' - \alpha a + \nu a^3 = 0. \tag{10.110}$$

This equation is the same as the equation for nonlinear oscillator. It has the integral

$$(a')^2 = \alpha a^2 - \frac{\nu}{2} a^4 + h, \tag{10.111}$$

where h is a constant. When $h = 0$ the trajectory lies on the separatrix in the phase plane, (a', a). Corresponding to this choce of parameters, the solution can be written in the form

$$a = \left(\frac{2\alpha}{\nu} \right) \frac{1}{\cosh(x - Vt)}. \tag{10.112}$$

This expression describes the solitary wave. Finaly we have the solution to equation (10.90):

$$u(x, t) = \left(\frac{2\alpha}{\nu} \right) \frac{\exp \left(i \left(\frac{V}{2} x - \left(\frac{V^2}{4} - \alpha \right) t \right) \right)}{\cosh(x - Vt)}. \tag{10.113}$$

Chapter 11

Inverse Scattering Method

Conservation laws for the KdV equation. Inverse scattering method for the KdV Equation. Gelfand-Levitan-Marchenko equation. N-soliton solutions. Meaning of the inverse scattering method. The Lax's L-A pair. Other equations solvable with inverse scattering method.

***

In 1967 Gardner C.S., Green J.M., Kruskal M.D., and Miura R.M. (*Gardner C.S., Green J.M., Kruskal M.D., and Miura R.M., Method for Solving Korteweg-De Vries Equation, Phys. Rev. Lett. 19, 1095 (1967)*) have invented a method of solution for the Korteweg de Vries equation. They have shown that to find a solution of the Cauchy problem for this nonlinear equation in partial derivatives it is sufficient to integrate two linear equations. One of them is a linear ordinary differential equation and the other is a linear integral equation. This method is called the *Inverse Scattering Method*. Then it has been found that there are many equations solvable with this method and among them there are the Nonlinear Schrödinger and the Sin-Gordon equations.

The canonical Form of the Korteweg de Vries Equation is given by (10.6). It is

$$\partial_t u - 6u\partial_x u + \partial_{xxx} u = 0. \tag{11.1}$$

Changing $u \to -\varepsilon u/6$ we can also write the KdV equation in the form

$$\partial_t u + \varepsilon u\partial_x u + \partial_{xxx} u = 0, \tag{11.2}$$

where parameter ε for small amplitude perturbations is supposed to be small, $\varepsilon \ll 1$. In this case the one-soliton solution (10.40) takes the form

$$u(x,t) = \frac{12\alpha^2}{\varepsilon \cosh^2(\alpha(x - 4\alpha^2 t))}. \tag{11.3}$$

Since the small parameter ε is in the denominator, it means that the soliton solution can not be obtained as a result of perturbation technics. To describe the solitons one must elaborate special technics.

11.1   Conservation Laws for the KdV Equation

In the case of the Burgers equation we observed that it has conserved integral $S = \int_{-\infty}^{+\infty} u(x,t)dt$, and the Cole-Hopf transformation $u = -2\nu\partial_x \ln z$ reduces the Cauchy problem for the Burgers equation to the Cauchy problem for the diffusion equation, which is a linear differential equation with constant coefficients.

The Korteweg de Vries equation has an infinite number of conservation laws, that is, an infinite number of integrals of motion. To demonstrate this we use the Gardner-Miura transformation of the function $u(x,t)$:

$$u = w + \varepsilon\partial_x w + \varepsilon^2 w^2, \tag{11.4}$$

where ε is a parameter. Substituting this expression into equation (11.1) we obtain

$$\left(1 + \varepsilon\partial_x + 2\varepsilon^2 w\right)\hat{\mathbf{R}}w = 0. \tag{11.5}$$

Here the operator $\hat{\mathbf{R}}$ acts on a function w as

$$\hat{\mathbf{R}}w \equiv \partial_t w - 6(w + \varepsilon^2 w^2)\partial_x w + \partial_{xxx}w. \tag{11.6}$$

If the function w obeys equation $\hat{\mathbf{R}}w = 0$ then the function u, given by relationship (11.4), is a solution to the KdV equation (11.1).

The function u as a solution of the KdV equation (11.1) does not depend on the parameter ε, that is why we may write a formal solution of the Gardner equation $\hat{\mathbf{R}}w = 0$ as a series

$$w = \sum_{n=0}^{\infty} \varepsilon^n w_n[u], \tag{11.7}$$

where $w_n = w_n[u]$ are given by the recurrence formulas

$$w_0 = u, \qquad w_1 = -\partial_x u, \qquad w_2 = \partial_{xx}u - u^2,\dots \quad . \tag{11.8}$$

Substituting expansion (11.7) into (11.5) we see that all the functions w_n obey the Gardner equation

$$\partial_t w_n + \partial_x(-3w_n^2 - 2\varepsilon^2 w_n^3 + \partial_{xx}w) = 0. \tag{11.9}$$

Assuming that w_n and its derivatives with respect to x tend to zero at infinity, and integrating this equation with respect to x, we obtain

$$\partial_t \int_{-\infty}^{+\infty} w_n[u]dx = 0, \qquad \text{that is} \qquad I_n = \int_{-\infty}^{+\infty} w_n[u]dx = \text{const} \tag{11.10}$$

for all $n = 0, 1, 2, \dots$. Thus, there is an infinite number of the integrals of motion I_n for the KdV equation. The firsts integrals are

$$I_0 = \int_{-\infty}^{+\infty} u(x,t)dx, \qquad I_2 = \int_{-\infty}^{+\infty} u^2(x,t)dx, \qquad I_4 = \int_{-\infty}^{+\infty} \left(u^3 + \tfrac{1}{2}(\partial_x u)^2\right)dx, \quad \dots$$

$$\tag{11.11}$$

11.2   Method of Inverse Scattering Problem to Solve the KdV Equation

We consider the transformation from the function $u(x,t)$ to $\psi(x,t)$ given by

$$u = \frac{\partial_{xx}\psi}{\psi} + \lambda, \tag{11.12}$$

where λ is a parameter. This expression can be rewritten in the form

$$\partial_{xx}\psi + (\lambda - u(x,t))\psi = 0, \tag{11.13}$$

which is the stationary Schrodinger equation with a potential $u(x,t)$ for the ψ-function. In this equation both the potential $u(x,t)$ and function $\psi(x,t)$ depend on t as on the parameter. For arbitrary potential $u(x,t)$ the eigenvalue of the Schrödinger equation λ must depend on time. However in the case when $u(x,t)$ is the solution of the KdV equation, λ does not depend on time. To show this we substitute (11.12) into (11.1) and obtain

$$\partial_t \lambda \psi^2 + \partial_x(\psi \partial_x Q - Q \partial_x \psi) = 0. \tag{11.14}$$

Here

$$Q = \partial_t \psi + \partial_{xxx}\psi - 3(u + \lambda)\partial_x \psi. \tag{11.15}$$

Using the boundary conditions $\lim_{x\to\pm\infty}\psi = 0$ and $\int_{-\infty}^{+\infty}\psi^2 dx \neq 0$, we obtain

$$\dot{\lambda} = 0, \tag{11.16}$$

and equation

$$\partial_t\psi + \partial_{xxx}\psi - 3(u + \lambda)\partial_x\psi = C\psi + D\psi \int \frac{dx}{\psi^2}, \tag{11.17}$$

from which we can find the time dependence of the ψ-function. Here C and D are the integration constants.

Now we discuss how to find localized, that is $\lim_{x\to\pm\infty} u(x,t) = 0$, solutions of the KdV equation for the given initial condition: $u(x,0) = u_0(x)$. In this case the integration constant D is equal to zero. As a first step we must solve equation (11.13) with $u = u_0(x)$. As a result we find $\psi(x,0)$, which gives the initial condition for equation (11.17), and the eigenvalues λ. Solving this equation we would obtain the ψ-function for all x and t and then find $u(x,t)$ from the Schrödinger equation (11.13). This is a very complicated problem because equation (11.17) contains an unknown function $u(x,t)$. However we do not need to know the ψ-function for all x and t: to restore a form of localized in the finite region potential in the Schrödinger equation it is sufficient to know the asymptotic behavior of the ψ-function at $|x| \to \infty$, where $u(x,t)$ vanishes, i.e. we would find a solution to **the inverse scattering problem**.

Solution of the inverse scattering problem is equivalent to the solution of a linear integral equation, so called the Gelfand-Levitan-Marchenko equation, with respect to function $K(x, y)$ of two independent variables:

$$K(x, y) + B(x + y) + \int_x^\infty B(y + z) K(x, z) dz = 0, \qquad (11.18)$$

where B is supposed to be given function. If equation (11.18) is solved with respect to $K(x, y)$ then the potential $u(x)$ is given by the relationship

$$u(x) = -2 \frac{d}{dx} K(x, x). \qquad (11.19)$$

The kernel in the Gelfand-Levitan-Marchenko equation is equal to

$$B(z) = \frac{1}{2\pi} \int_{-\infty}^{+\infty} b(k) \exp(ikz) + \sum_{n=1}^N c_n^2 \exp(\kappa_n z), \qquad (11.20)$$

where the first termin the right hand side is due a contribution of the continous spectrum and the second one is due to the discret spectrum..

Depending on the specific form of the potential $u_0(x)$ the spectrum of equation (11.17) contains discrete ($\lambda < 0$) and continuous ($\lambda > 0$) parts.

For the discrete spectrum we have

$$\lambda_n = -\kappa_n^2, \qquad n = 1, 2, \ldots \qquad (11.21)$$

with eigenfunctions ψ_n such that $\lim_{|x| \to \infty} |\psi_n| = 0$, and $0 < \int_{-\infty}^{+\infty} \psi_n^2 dx < \infty$. If we normalize the eigenfunctions ψ_n to have $\int_{-\infty}^{+\infty} \psi_n^2 dx = 1$, then asymptotically at $|x| \to \infty$ the eigenfunctions ψ_n are equal to

$$\psi_n = c_n \exp(-\kappa_n |x|), \qquad (11.22)$$

with c_n being the normalization constants.

In the case of the continuous spectrum ($\lambda > 0$), the asymptotic behavior of the ψ−functions is

$$\begin{array}{llll}
\psi(x, t) & \sim & \exp(-ikx) + b(k, t) \exp(ikx) & \text{at } x \to +\infty, \\
\psi(x, t) & \sim & a(k, t) \exp(-ikx) & \text{at } x \to -\infty
\end{array} \qquad (11.23)$$

with $k^2 = \lambda$. The coefficients $a(k, t)$ and $b(k, t)$ are considered here as the transmission and reflection amplitudes with $|a(k, t)|^2 + |b(k, t)|^2 = 1$. This is illustrated in Fig. 11.1.

Thus, we have to know the time dependence of the functions κ_n, c_n, a_n, and b_n. We have seen that κ_n does not depend on time. To find c_n, a_n, and b_n we need to solve equation (11.17) with $D = 0$.

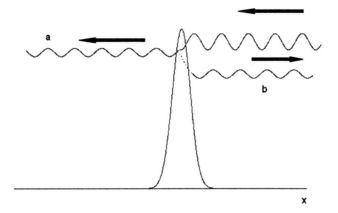

Fig. 11.1. Transmission and reflection on the potential $u(x)$.

In the case of the discrete spectrum we use a condition $\lim_{|x|\to\infty} |\psi_n| = 0$. Multiplying equation (11.17) by ψ_n and integrating on x we obtain

$$\frac{1}{2}\partial_t \int_{-\infty}^{+\infty} \psi_n^2 dx + \int_{-\infty}^{+\infty} \partial_x \left(\psi_n \partial_{xx}\psi_n - 2(\partial_x\psi_n)^2 - 3\lambda\psi_n^2\right) dx = C \int_{-\infty}^{+\infty} \psi_n^2 dx. \tag{11.24}$$

Since $\int_{-\infty}^{+\infty} \psi_n^2 dx = 1$ and $\lim_{|x|\to\infty} |\psi_n| = 0$ we have $C = 0$. For $c_n(t)$ it gives equation

$$\dot{c}_n - 4\kappa_n^3 c_n = 0, \tag{11.25}$$

that is

$$c_n(t) = c_n(0) \exp(4\kappa_n^3 t), \qquad \kappa_n^2 = -\lambda. \tag{11.26}$$

Substituting asymptotic expressions (11.23) into equation (11.17) we find

$$(\dot{b} - (4ik^3 + C)b) \exp(ikx) + (-C + 4ik^3) \exp(-ikx) = 0, \tag{11.27}$$

and

$$(\dot{a} + (4ik^3 - C)a) \exp(-ikx) = 0. \tag{11.28}$$

By virtue of independence of the functions $\exp(ikx)$ and $\exp(-ikx)$ we have $C = 4ik^3$ and

$$\begin{aligned} a(k,t) &= a(k,0), \\ b(k,t) &= b(k,0) \exp(8ik^3 t), \quad k^2 = \lambda > 0. \end{aligned} \tag{11.29}$$

Now we know the scattering data, which are sufficient to find the potential $u(x,t)$.

11.3 N-Soliton Solution

We consider the initial condition for KdV equation in the form

$$u(x,0) = -\frac{2}{\cosh^2 x}. \tag{11.30}$$

The discrete spectrum has just one eigenvalue, $\lambda = -\kappa_1^2 = 1$. The eigenfunction at $t = 0$ is $\psi_1(x,0) = 1/(2\cosh(x))$ with $c_1(0) = \sqrt{2}$. In the continuum part of the spectrum we have $b(k,0) = 0$, because this form of the potential is known in the quantum mechanics as the reflectionless potential. Calculating the kernel of the Gelfand-Levitan-Marchenko equation according to equation (11.20), we find

$$B(x+y,t) = 2\exp(-8t + (x+y)). \tag{11.31}$$

Substituting it into equation (11.18) we obtain

$$K(x,y,t) + 2\exp(-8t + (x+y)) - 2\exp(-8t + y)\int_{-\infty}^{x} \exp(z)K(x,z,t)dz = 0. \tag{11.32}$$

Now setting $K(x,y,t) = L(x)\exp(y)$ we have

$$(1 + \exp(-8t + 2x))L(x) + 2\exp(-8t + 2x) = 0. \tag{11.33}$$

We finally obtain $\partial_t K(x,x,t) = -2/(1 + \exp(-8t + 2x))$ and

$$u(x,t) = -\frac{2}{\cosh^2(x - 4t)}. \tag{11.34}$$

This is the one-soliton solution of the KdV equation.

For more deeper potential than (11.30) as it is given by expression

$$u(x,0) = -\frac{6}{\cosh^2 x} \tag{11.35}$$

we have, in the discrete spectrum,

$$
\begin{aligned}
\lambda_1 &= -\kappa_1^2 = -4, & \psi_1 &= \frac{\sqrt{3}}{2\cosh^2 x}, \\
\lambda_2 &= -\kappa_2^2 = -1, & \psi_2 &= \sqrt{\frac{3\tanh x}{2\cosh x}}.
\end{aligned}
\tag{11.36}
$$

The form of the potential is chosen to be reflectionless, that is $b(k,0) = 0$. The solution of the Gelfand-Levitan-Marchenko equation in this case gives

$$u(x,t) = -12\left(\frac{3 + 4\cosh(2x - 8t) - \cosh(4x - 64t)}{(\cosh(3x - 36t) + 3\cosh(x - 28t))^2}\right). \tag{11.37}$$

This is the two-soliton solution of the KdV equation.

When $x \to -\infty$ we have the first soliton with $u_1(x,t) \approx -8/\cosh^2(2x - 32t)$ and the second with $u_2(x,t) \approx -2/\cosh^2(x - 4t + s)$, where the soliton phase is equal to $s = (\ln 3)/2$.

In the case of an N-soliton solution the Gelfand-Levitan-Marchenko equation (11.18) has a form

$$K(x,y,t) + \sum_{n=1}^{N} c_n^2(t)\left(\exp(-\kappa_n(x+y)) + \int_{x}^{+\infty} \exp(-\kappa_n(z+y))K(x,z,t)dz\right) = 0. \tag{11.38}$$

Setting

$$K(x, y, t) = -\sum_{n=1}^{N} c_n(t)\varphi_m(x, t)\exp(-\kappa_n y) \qquad (11.39)$$

and substituting this expression into equation (11.38) we obtain a system of linear algebraic equations for functions $\varphi_m(x, t)$. It reads

$$\sum_{n=1}^{N}\left(\delta_{mn} + c_m(t)c_n(t)\,\frac{\exp(-(\kappa_n + \kappa_m)x)}{\kappa_n + \kappa_m}\right)\varphi_n(x, t) = c_m(t)\exp(-\kappa_n x),$$
$$m = 1, 2, \ldots, N.$$
$$(11.40)$$

Solving this system in accordance with Krammer's rule we obtain

$$\varphi_n(x, t) = \frac{1}{\mathcal{D}}\sum_{m=1}^{N} c_m(t)\exp(-\kappa_m x)Q_{mn}(x, t), \qquad (11.41)$$

where Q_{mn} and $\mathcal{D}$ is the cofactor and determinant of the matrix F_{mn} :

$$F_{mn} = \delta_{mn} + c_m(t)c_n(t)\frac{\exp(-(\kappa_n + \kappa_m)x)}{\kappa_n + \kappa_m}. \qquad (11.42)$$

Now we can calculate the function $K(x, x, t)$. It is

$$\begin{aligned}
K(x, x, t) &= -\sum_{m=1}^{N} c_m(t)\varphi_m(x, t)\exp(-\kappa_m x) \\
&= -\frac{1}{\mathcal{D}}\sum_{m=1}^{N}\sum_{n=1}^{N} c_m(t)c_n(t)\exp(-(\kappa_n + \kappa_m)x)Q_{mn}(x, t) \\
&= \frac{1}{\mathcal{D}}\frac{d}{dx}\mathcal{D} = \frac{d}{dx}\ln|\mathcal{D}|.
\end{aligned}$$
$$(11.43)$$

This formula recovers the results obtained above with the Hirota transformation. It also gives a generic expression for the reflectionless potentials for the Schrödinger equation, which can be parametrized by N arbitrary numbers $c_m(t)$ and corresponding eigenvalues $\lambda_m = -\kappa_m^2$, $m = 1, 2, \ldots, N$. On the other hand, formula (11.43) describes solutions of the KdV equation which correspond to N interacting solitons with their amplitudes $-2\kappa_m^2$ and velocities of propagation $4\kappa_m^2$.

11.3.1 Meaning of the Inverse Scattering Method. The Lax's L-A Pair.

We consider the one-parametric family of the Schrödinger operators

$$\hat{\mathbf{L}} = \hat{\mathbf{D}}^2 - u(x, t), \qquad \hat{\mathbf{D}} = \partial_x, \qquad (11.44)$$

which depends on the parameter t.

There is a question, what kind of constraints must be imposed on function $u(x,t)$ for all the operators of this family to have the same spectrum, particularly, the same eigenvalues $\lambda_m(t) \equiv \lambda_m(0)$. It is known that the necessary and sufficient condition for that is for the operators $\hat{L}(t)$ to be unitary equivalent to the operator $\hat{L}(0) = \hat{D}^2 - u(x,0)$. It means there is such a unitary operator $\hat{U}(t)$ that

$$\hat{L}(0) = \hat{U}^*(t)\hat{L}(t)\hat{U}(t), \tag{11.45}$$

where $\hat{U}^*(t)$ is an operator conjugated to $\hat{U}(t)$.

Now we discuss the properties which the potential $u(x,t)$ must have for the spectrum of the operator $\hat{L}(t)$ to be invariant. Here we follow *P. D. Lax, Integrals of nonlinear equations of evolution and solitary waves. Com. Pure Appl. Math. 21, 467 (1968)*.

We introduce an operator

$$\hat{A} = \hat{U}_t\hat{U}^*, \tag{11.46}$$

where $\hat{U}_t$ is a derivative of the operator $\hat{U}$ with respect to the parameter t. Using $\hat{A}^* = (\hat{U}_t\hat{U}^*)^* = \hat{U}\hat{U}_t^*$ and $\hat{U}\hat{U}^* = E$, where E is the unity operator, we obtain

$$\hat{A} + \hat{A}^* = \hat{U}_t\hat{U}^* + \hat{U}\hat{U}_t^* = (\hat{U}\hat{U}^*)_t = 0. \tag{11.47}$$

We see that $\hat{A}$ is the antisymmetric operator: $\hat{A} = -\hat{A}^*$.

Multiplying equation $\hat{A} = \hat{U}_t\hat{U}^*$ on the operator $\hat{U}$ from the right hand side, we find

$$\hat{A}\hat{U} = \hat{U}_t\hat{U}^*\hat{U} = \hat{U}_t, \quad \text{i.e.} \quad \hat{U}_t = \hat{A}\hat{U}. \tag{11.48}$$

This is a differential equation for the operator $\hat{U}$, which must be solved with initial condition $\hat{U}(0) = E$. Now we need to find the equation for the operator $\hat{A}$.

Differentiating equation (11.45) with respect to time we obtain

$$\hat{U}_t^*(t)\hat{L}(t)\hat{U}(t) + \hat{U}^*(t)\hat{L}_t(t)\hat{U}(t) + \hat{U}^*(t)\hat{L}(t)\hat{U}_t(t) = 0. \tag{11.49}$$

Using equation (11.48) and $\hat{U}_t^* = -\hat{U}^*\hat{A}$, we rewrite the equation (11.49) in the form

$$-\hat{U}^*\hat{A}\hat{L}\hat{U} + \hat{U}^*\hat{L}_t\hat{U} + \hat{U}^*\hat{L}\hat{A}\hat{U} = 0. \tag{11.50}$$

Multiplying this equation from the right-hand side by $\hat{U}^*$, and from the left-hand side by $\hat{U}$, we obtain

$$\hat{L}_t - \hat{A}\hat{L} + \hat{L}\hat{A} = 0 \tag{11.51}$$

or

$$\hat{L}_t + \left[\hat{L}, \hat{A}\right] = 0, \tag{11.52}$$

where $[\ldots, \ldots]$ denotes a commutator of operators.

Then, we see that

$$\hat{L}_t = -u_t. \tag{11.53}$$

That is

$$u_t = \left[\hat{\mathbf{L}}, \hat{\mathbf{A}}\right],\tag{11.54}$$

and $\left[\hat{\mathbf{L}}, \hat{\mathbf{A}}\right]$ must be a multiplication operator.

We have found the necessary conditions under which the spectrum of the operator $\hat{\mathbf{L}}(t)$ is invariant. The operator $\left[\hat{\mathbf{L}}, \hat{\mathbf{A}}\right]$ must be a multiplication operator and the differential equation for $\hat{\mathbf{L}}(t)$ is equation (11.52). We can prove that the conditions are also sufficient for the spectrum of the operator $\hat{\mathbf{L}}(t)$ to be invariant.

Let us assume that there is an operator $\hat{\mathbf{A}}$ with discussed above properties. Thence the operator $\hat{\mathbf{U}}(t)$ is also known. We consider an operator

$$\hat{\mathbf{L}}(t) = \hat{\mathbf{U}}^*(t)\hat{\mathbf{L}}(t)\hat{\mathbf{U}}(t).\tag{11.55}$$

Differentiating this operator with respect to time, we obtain

$$\frac{d}{dt}\hat{\mathbf{L}}(t) = \hat{\mathbf{U}}^*(t)(\hat{\mathbf{L}}_t(t)+\left[\hat{\mathbf{L}}, \hat{\mathbf{A}}\right])\hat{\mathbf{U}}(t),\tag{11.56}$$

that is $\hat{\mathbf{L}}_t(t) = 0$ and $\hat{\mathbf{L}}(0) = \hat{\mathbf{L}}(0) = \hat{\mathbf{L}}(t)$. This is equivalent to the sufficient condition of the invariance of the spectrum of the operator $\hat{\mathbf{L}}(t)$.

Assuming that for the potential $u(x,t)$ the spectrum of the operator $\hat{\mathbf{L}}(t)$ is invariant and the operator $\hat{\mathbf{A}}$ exists, we obtain an equation which describes the ψ−function evolution. Dependence of the ψ−function on the x−coordinate must be found from equation

$$\hat{\mathbf{L}}(t)\psi(t) + \lambda\psi(t) = 0.\tag{11.57}$$

Since at $t = 0$

$$\hat{\mathbf{L}}(0)\psi(0) + \lambda\psi(0) = 0, \qquad \psi(0) = \psi(x,0),\tag{11.58}$$

according to equation(11.45) we have

$$\hat{\mathbf{U}}^*(t)\hat{\mathbf{L}}(t)\hat{\mathbf{U}}(t)\psi(0) + \lambda\psi(0) = 0.\tag{11.59}$$

Multiplying this relationship by $\hat{\mathbf{U}}(t)$ from the left-hand side, we obtain

$$\hat{\mathbf{L}}(t)\hat{\mathbf{U}}(t)\psi(0) + \lambda\hat{\mathbf{U}}(t)\psi(0) = 0,\tag{11.60}$$

from which it follows that

$$\psi(t) = \hat{\mathbf{U}}(t)\psi(0).\tag{11.61}$$

Since

$$\psi_t(t) = \hat{\mathbf{U}}_t(t)\psi(0) = \hat{\mathbf{A}}(t)\hat{\mathbf{U}}(t)\psi(0) = \hat{\mathbf{A}}(t)\psi(t),\tag{11.62}$$

we have equation

$$\psi_t(t) = \hat{\mathbf{A}}(t)\psi(t),\tag{11.63}$$

that describes the time dependence of the ψ-function.

The operators $\hat{\mathbf{L}}$ and $\hat{\mathbf{A}}$ are called "the Lax's $\hat{\mathbf{L}}$-$\hat{\mathbf{A}}$ pair of operators".

Let us assume that an operator $\hat{\mathbf{A}}$ is

$$\hat{\mathbf{A}} = c\hat{\mathbf{D}} \tag{11.64}$$

with constant c, then

$$[\hat{\mathbf{L}}, \hat{\mathbf{A}}]\varphi = c(\hat{\mathbf{D}}^3\varphi - u\hat{\mathbf{D}}\varphi - \hat{\mathbf{D}}^3\varphi + \hat{\mathbf{D}}(u\varphi)) = cu_x\varphi, \tag{11.65}$$

where φ is some function on which the operator acts. We have found that the commutator $[\hat{\mathbf{L}}, \hat{\mathbf{A}}]$ for $\hat{\mathbf{A}} = c\hat{\mathbf{D}}$ is the operator of multiplication on $c\hat{\mathbf{U}}_x$ and equation (11.54) takes the form of the transport equation

$$u_t - cu_x = 0, \tag{11.66}$$

with c being a speed of the wave propagation. Thus, in the case $u(x,t) = f(x + ct)$ the spectrum of the operator $\hat{\mathbf{L}}(t) = \hat{\mathbf{D}}^2 - u(x,t)$ does not depend on time. However, this result is apparent because the operator $\hat{\mathbf{L}}(t)$ with $u(x,t) = f(x + ct)$ is invariant under the change of independent variable $x \to x + ct$.

We obtain a more important result by considering

$$\hat{\mathbf{A}} = -4\hat{\mathbf{D}}^3 + 6u\hat{\mathbf{D}} + 3u_x. \tag{11.67}$$

Calculating the commutator $[\hat{\mathbf{L}}, \hat{\mathbf{A}}]$ we find

$$[\hat{\mathbf{L}}, \hat{\mathbf{A}}] = 6uu_x - u_{xxx}. \tag{11.68}$$

It is a multiplication operator, thence equation (11.52) is the KdV equation. We see that if $u(x,t)$ is a solution of the KdV equation, the spectrum of the operator $\hat{\mathbf{L}}(t) = \hat{\mathbf{D}}^2 - u(x,t)$ does not depend on time. In this case the time evolution of the ψ-function is described by equation (11.63), which we write as

$$\psi_t = -4\psi_{xxx} + 6u\psi_x + 3u_x\psi. \tag{11.69}$$

Using the equation $\psi_{xx} - (u - \lambda)u = 0$, we can transform (11.69) to (11.17).

11.4   Other Equations Solvable with Inverse Scattering Method

The nonlinear Schrödinger equation

$$i\partial_t u + \partial_{xx} u + \nu|u|^2 u = 0 \tag{11.70}$$

according to *V. E. Zakharov and A. B. Shabat, 1971-1974* can be described by the $\hat{\mathbf{L}}$ and $\hat{\mathbf{A}}$ operators as $\hat{\mathbf{L}}_t + \left[\hat{\mathbf{L}}, \hat{\mathbf{A}}\right] = 0$ with modified equations for the inverse scattering problem. In this case the operators $\hat{L}$ and $\hat{A}$ are matrix differential operators

$$\hat{L} = \begin{pmatrix} 1+p & 0 \\ 0 & 1-p \end{pmatrix} \hat{D} - \begin{pmatrix} 0 & iu* \\ iu & 0 \end{pmatrix}, \tag{11.71}$$

$$\hat{A} = \begin{pmatrix} -p & 0 \\ 0 & -p \end{pmatrix} \hat{D}^2 - \begin{pmatrix} |u|^2/(1+p) & i\partial_x u* \\ -i\partial_x u & -|u|^2/(1-p) \end{pmatrix}, \tag{11.72}$$

where $\nu = 2/(1-p^2)$.

The sine-Gordon equation

$$\partial_{tt} u - \partial_{xx} u = \sin u \tag{11.73}$$

is related to the matrix operator

$$\hat{L} = \begin{pmatrix} i & 0 \\ 0 & -i \end{pmatrix} \hat{D} + \frac{1}{2} \begin{pmatrix} 0 & i\partial_x u* \\ i\partial_x u & 0 \end{pmatrix}, \tag{11.74}$$

and the equation for the evolution of eigenfunctions

$$\partial_t \begin{pmatrix} \psi_1 \\ \psi_2 \end{pmatrix} = \frac{i}{4\lambda} \begin{pmatrix} \cos u & \sin u \\ \sin u & -\cos u \end{pmatrix} \begin{pmatrix} \psi_1 \\ \psi_2 \end{pmatrix} \tag{11.75}$$

with λ being the spectral parameter in the scattering problem.

Chapter 12

Bäcklund Transformation

General properties. The Cole-Hopf solution as the Bäcklund transformation. The Bäcklund transformation for Liouville equation. Auto-Bäcklund transformation. Bäcklund transformation of the sin-Gordon equation. Bäcklund transformation of the KdV equation.

***

12.1 General Properties

The Bäcklund transformation can be used to connect solutions of a nonlinear equation either to those of a nonlinear equation whose properties are better understood, or to those of a linear equation. The Hopf-Cole transformation of the Burgers equation provides an example of this. Alternatively, the invariance of the equation under the Bäcklund transformation means that a given solution of the equation is transformed into another solution of the same equation. This property can be used to generate hierarchies of solutions to nonlinear equations by algebraic methods, so producing the nonlinear superposition principle.

The Bäcklund transformation gives the relationship between two surfaces described by the parameters

$$x, \; y, \; u, \partial_x u, \partial_y u \tag{12.1}$$

and

$$x', \; y', \; u', \; \partial_{x'} u', \partial_{y'} u'. \tag{12.2}$$

These quantities are related by the total differentials

$$du = \partial_x u \, dx + \partial_y u \, dy, \tag{12.3}$$

and

$$du' = \partial_{x'} u' \, dx' + \partial_{y'} u' \, dy', \tag{12.4}$$

respectively. So that only four independent equations are required to map between primed and unprimed coordinate systems:

$$F_j\left(x, y, u, \partial_x u, \partial_y u; x', y', u', \partial_{x'} u', \partial_{y'} u'\right) = 0 \qquad (12.5)$$

for $j = 1, 2, 3,$ and 4. It gives a mapping between two second-order differential
equations. These four equations are called the Bäcklund transformation. Here
x, y are the coordinates and $u(x, y)$ is the dependent variable.

The frequently used form of equation (12.5) is

$$\partial_x u = F_1\left(x', y', u', \partial_{x'} u', \partial_{y'} u'; u\right),$$

$$\partial_y u = F_2\left(x', y', u', \partial_{x'} u', \partial_{y'} u'; u\right),$$

$$x = F_3\left(x', y', u', \partial_{x'} u', \partial_{y'} u'; u\right),$$

$$y = F_4\left(x', y', u', \partial_{x'} u', \partial_{y'} u'; u\right) \qquad (12.6)$$

with similar expressions for primed coordinate system.

Taking into account relationships for the differentials (12.3) and (12.4) we
obtain that these expressions yeld the following dependences

$$du = \partial_x u\, dx + \partial_y u\, dy = F_1 dx + F_2 dy, \qquad (12.7)$$

$$du' = \partial_{x'} u'\, dx' + \partial_{y'} u'\, dy' = F_1' dx' + F_2' dy'. \qquad (12.8)$$

For these expressions to be exact differentials the mixed second partial derivatives

$$\partial_y F_1 = \partial_x F_2, \qquad \text{and} \qquad \partial_{y'} F_1' = \partial_{x'} F_2' \qquad (12.9)$$

must be equal, or it is equivalent to

$$\partial_{xy} u = \partial_{yx} u, \qquad \text{and} \qquad \partial_{x'y'} u' = \partial_{y'x'} u'. \qquad (12.10)$$

12.1.1   The Cole-Hopf Solution as the Bäcklund Transformation

The Cole-Hopf transformation of the Burgers equation to the diffusion equation
can be written as a Bäcklund transformation.

Equations $u(x, t) = -\partial_x \phi$ and $z(x, t) = \exp\left(\psi(x, t)/2\right)$ with $\phi(x, t) = \psi(x, t) + \int_0^t f(s) ds$ and $f(t)$ being a function of integration, we can rewrite
as

$$\partial_x z = -\frac{1}{2} u z. \qquad (12.11)$$

The solution to the diffusion equation $\partial_t z - \partial_{xx} z = 0$, $z(x, t)$, can be obtained
from equation (12.11) :

$$z(x, t) = \exp\left(-\frac{1}{2} \int^x u(s, t) ds\right). \qquad (12.12)$$

Differentiating both sides of the equation with respect to time and using the fact that $u(x,t)$ is a solution to the Burgers equation, we find

$$\partial_t z = \frac{1}{4}\left(u^2 - 2\partial_x u\right)z. \tag{12.13}$$

Equations (12.11) and (12.13) together with $x = x'$ and $t = t'$ constitute the Bäcklund transformation, relating the solutions of the diffusion equation to those of the Burgers equation. The integrability conditions (12.10) show that $u(x,t)$ is a solution to the Burgers equation.

12.1.2   The Bäcklund Transformation for Liouvile Equation

We suppose that a nonlinear differential equation has the form

$$\partial_{tx} u = \exp(u), \tag{12.14}$$

then the Bäcklund transformation becomes

$$\partial_{x'} u' = \partial_x u + \lambda \exp\left(\frac{1}{2}(u + u')\right),$$

$$\partial_{y'} u' = -\partial_y u - \frac{1}{\lambda}\exp\left(\frac{1}{2}(u - u')\right), \tag{12.15}$$

with $x' = x$, and $y' = y$. Here the quantity λ is a real constant that is called the Bäcklund parameter and represents a degree of freedom of the transformation. The integrability condition for these equations gives

$$\partial_{x'y'} u' = 0. \tag{12.16}$$

12.1.3   Auto-Bäcklund Transformation

This transformaton leaves an equation invariant. The invariance does not imply that each solution is transformed into itself.

12.1.4   Bäcklund Transformation of the Sin-Gordon Equation

We consider the Bäcklund transformation of the form

$$\partial_{x'} u' = \partial_x u - 2\lambda \sin\left(\frac{1}{2}(u + u')\right),$$

$$\partial_{y'} u' = -\partial_y u - \frac{2}{\lambda}\sin\left(\frac{1}{2}(u - u')\right), \tag{12.17}$$

where $x' = x$, $y' = y$ and λ is a real constant.

Differentiating $\partial_{x'}u'$ with respect to y' and $\partial_{y'}u'$ with respect to x' and using the integrability conditions we find, eliminating u from equations, that u' obeys the equation

$$\partial_{x'y'}u' = \sin u', \tag{12.18}$$

which is the sine-Gordon equation. We can find the same equation for u :

$$\partial_{xy}u = \sin u. \tag{12.19}$$

This is an auto-Bäcklund transformation. From the first two equations (12.17) we can obtain a quasi-linear first-order equation

$$\frac{1}{\lambda}\partial_x u - \lambda \partial_y u = \frac{1}{\lambda}\partial_{x'}u' + \lambda \partial_{y'}u' + 4\sin\left(\frac{1}{2}u\right)\cos\left(\frac{1}{2}u'\right). \tag{12.20}$$

If $u'(x', y')$ is a known soliton of the sine-Gordon equation, the function $u(x, y)$ can be found by integrating equation (12.20) with the charachteristic methods.

If we take $u' \equiv 0$ to be a trivial solution to the sin-Gordon equation, equations (12.17) become

$$\partial_x u = 2\lambda \sin\left(\frac{u}{2}\right),$$

$$\partial_y u = \frac{2}{\lambda}\sin\left(\frac{u}{2}\right), \tag{12.21}$$

meanwhile (12.20) takes the form

$$\partial_x u - \lambda^2 \partial_y u = 0. \tag{12.22}$$

Seeking its solution in the form $u = f(\lambda x + y/\lambda)$ and substituting it into equations (12.21) we obtain the equation for the function $f(\xi)$:

$$\partial_\xi f = 2\sin\left(\frac{f}{2}\right). \tag{12.23}$$

Integrating this equation we find a solution

$$u(x, y) = 4\arctan\left(\exp\left(\lambda x + \frac{y}{\lambda}\right)\right), \tag{12.24}$$

which corresponds to the soliton solution of the sin-Gordon equation.

If we then substitute the one soliton solution into equation (12.20) instead of u', and integrate it, we obtain

$$u(x, y) = 4\arctan\left(\frac{\lambda^2 - 1}{\lambda^2 + 1}\frac{\sinh\left(\frac{x+y}{2}\left(\lambda + \frac{1}{\lambda}\right)\right)}{\cosh\left(\frac{x-y}{2}\left(\lambda - \frac{1}{\lambda}\right)\right)}\right). \tag{12.25}$$

This expression describes two interacting solitons.

12.1.5 Bäcklund Transformation of the KdV Equation

In the case of the KdV equation we set

$$\partial_x w = u, \tag{12.26}$$

where $u(x,t)$ is a solution of the KdV equation

$$\partial_t u - 6u\partial_x u + \partial_{xxx} u = 0. \tag{12.27}$$

We consider the transformation $w \to w'$ given by

$$\partial_{x'} w' = -\partial_x w - \lambda + \frac{1}{2}(w - w')^2,$$

$$\partial_{t'} w' = -\partial_t w - (w - w')(\partial_{xx} w - \partial_{x'x'} w') + 2((\partial_x w)^2 + \partial_x w \partial_{x'} w' + (\partial_{x'} w')^2), \tag{12.28}$$

where $x' = x$, $y' = y$ and λ is arbitrary constant. It is easy to show that if $\partial_x w = u$ is a solution to the KdV equation then $\partial_{x'} w' = u'$ also is the solution to the KdV equation.

Chapter 13

Self-Similar Solutions

Self-similar solutions. Problem of strong explosion. Self-similar solutions of the second type. Uniformly propagating waves in the framework of the Burgers and Korteweg-de Vries equations as self-similar solutions of the second type. Π-theorem. Self-similar solutions of gas dynamics equations. Self-similar solutions with homogeneous deformation.

***

Above we discussed several examples of self-similar solutions. They are of a particular type in which a dependent variable may be written as

$$f(x,t) = \varphi(t) F\left(\frac{x}{\lambda(t)}\right). \tag{13.1}$$

Here x, t are independent variables that are the spatial and time coordinates, and $\varphi(t)$ and $\lambda(t)$ provide the scales of dependent variable and spatial coordinate. These solutions are called *self-similar*, because the spatial dependence of the motions remains similar to itself. We see that the function $f(x,t)/\varphi(t)$ depends on one self-similar coordinate

$$\xi = \frac{x}{\lambda(t)}. \tag{13.2}$$

Hence the solution of nonlinear equations in partial derivatives is reduced to the solution of ordinary differential equations which is thought to be much simpler and might make it possible to obtain a number of exact particular solutions. The self-similar solutions correspond to a degenerate problem with the number of parameters less than the maximum number, which may be only in the case of the problems with a symmetry. Formally, it corresponds to the case when all parameters in initial and boundary conditions that have the dimensions of independent variables vanish or are infinite. However, a much more important property of the self-similar solutions is that which describes the intermediate asymptotic (*G. I. Barenblatt and Ya. B. Zel'dovich, 1972*) of the motion when all the specific details of the initial and boundary conditions do not play any significant role.

If the self-similarity of the motion is allowed, in many cases it may be estab-
lished from the dimensional analysis. Within more sophisticated and rigorous
approach the self-similar solutions can be obtained with the help of the Lie
group theory. These are so-called self-similar solutions of the first kind which
imply some regularity in the passage from initial non-self-similar regime of the
motion to the intermediate self-similar asymptotic regime. One of the most important examples of such regimes is provided by the Sedov and Taylor solution
that describes the blast wave produced by a point explosion. However, there are
self-similar solutions of the second kind, for which the dimensional analysis has
to be supplemented by a solution of some eigenvalue problem. The second kind
of self-similar motion can be seen in the Landau-Stanyukovich and Guderley
solution of the implosion problem.

13.1   Problem of Strong Explosion

An example of self-similar motion is provided by the gas dynamics at the initial
stage of a nuclear explosion as well as explosions of super nova stars.

After the very initial stage characterized by dominant role of the heat con-
duction and energy transport determined by radiation there begins the second
stage when the gas motion can be assumed to be adiabatic. At this stage the
shock wave overtakes the heat wave.

The gas motion due to a strong explosion is determined by the following
parameters: the amount of energy E released during explosion; the density
value in unperturbed gas that equals ρ_0, the pressure p_0 in unperturbed gas.
As well as the gas motion depends on time t, and radius r.

With those parameters we can find typical length and time scales. The
length scale can be constructed either to be equal to

$$L_1 = (E/p_0)^{1/2}, \quad \text{or to} \quad L_2 = R_0, \tag{13.3}$$

where R_0 is the size of the region the energy of the explosion released. The time
scales are

$$T_1 = (\rho_0 E^{2/3}/p_0^{5/3})^{1/2} \quad \text{and} \quad T_2 = (\rho_0 R_0^5/E)^{1/2}. \tag{13.4}$$

There is a correspondence: when $L_1 \gg L_2$ we have $T_1 \gg T_2$.

The region of parameters where the intermediate asymptotics, which have
been studied by *Sedov (1946) and Taylor (1950)*, are valid corresponds to

$$L_1 \gg r \gg L_2 \quad \text{and} \quad T_1 \gg t \gg T_2. \tag{13.5}$$

For the characteristics of motion (velocity, v, density, ρ, and pressure, p) we
have the representation

$$v = \frac{r}{t}V(\xi), \qquad \rho = \rho_0\Phi(\xi), \qquad p = p_0\left(\frac{r}{t}\right)^2 P(\xi), \tag{13.6}$$

with the self-similar variable

$$\xi = \frac{r}{k(Et^2/\rho_0)^{1/5}}. \tag{13.7}$$

We see the shock wave position at $\xi = 1$, which corresponds to its radius as a function of time, to be

$$R(t) = k(E/\rho_0)^{1/5}t^{2/5}. \tag{13.8}$$

Here a dimensionless number k is fixed from the definition of E in the flow

$$E = \int_0^{R(t)} \rho \left(\frac{1}{\gamma - 1} \frac{p}{\rho} + \frac{v^2}{2} \right) r^2 dr, \tag{13.9}$$

which leads to

$$4\pi k^5 \int_0^1 \Phi(\xi) \left(\frac{1}{\gamma - 1} \frac{P(\xi)}{\Phi(\xi)} + \frac{V^2(\xi)}{2} \right) \xi^2 d\xi = const. \tag{13.10}$$

For functions $V(\xi)$, $\Phi(\xi)$, and $P(\xi)$ from equations of gas dynamics

$$\partial_t \rho + \partial_r(\rho v) + \frac{n-1}{r}\rho v = 0, \tag{13.11}$$

$$\partial_t v + v\partial_r v = -\frac{1}{\rho}\partial_r p, \tag{13.12}$$

$$\partial_t \left(\frac{p}{\rho^\gamma} \right) + v\partial_r \left(\frac{p}{\rho^\gamma} \right) = 0 \tag{13.13}$$

with the boundary conditions at the strong shock wave front

$$v = \frac{2}{\gamma + 1}U, \qquad \rho = \frac{\gamma + 1}{\gamma - 1}\rho_0, \qquad p = \frac{2}{\gamma + 1}\rho_0 U^2 \tag{13.14}$$

behind the shock moving with velocity U we obtain ordinary differential equations. Here the space dimension $n = 1, 2, 3$ is assumed to be $n = 3$. Their solution describes strong explosion.

This is a self-similar solution of the first type.

13.2   Self-similar Solutions of the Second Type

We can modify the problem on strong explosion to get the self-similar solution of the second type. The modification consists of assuming the adiabat index γ behind the shock wave to be different from the parameter γ_1 that enters the condition at the shock wave front in the expression for the energy: $E = p(\gamma_1 - 1)\rho$. The fact that the inequality $\gamma_1 \neq \gamma$ corresponds to losses of energy (for $\gamma_1 < \gamma$ it is due to radiation losses proportional to $T^{3/2}$) or when $\gamma_1 > \gamma$ to the energy release at the shock wave front, means that the energy inside the region bounded by the shock wave is not a constant.

Substituting expressions (13.6) into the gas dynamics equations (13.11)-(13.13) and boundary conditions at the shock front, we come to contradiction, so that there exists no self-similar solution in the form (13.6) that satisfies all the conditions with $\gamma_1 \neq \gamma$. This is because the energy condition at the shock wave, being written in the form

$$\rho(v - U)\left(\frac{1}{\gamma - 1}\frac{p}{\rho} + \frac{v^2}{2}\right) + pv = \left(\frac{1}{\gamma - 1} - \frac{1}{\gamma_1 - 1}\right)\frac{p}{\rho}\rho(v - U) \qquad (13.15)$$

, shows that the energy (13.9) is no longer constant:

$$\dot{E} = -4\pi R^2 U\left(\frac{1}{\gamma - 1} - \frac{1}{\gamma_1 - 1}\right)\rho_0\frac{p}{\rho} \neq 0. \qquad (13.16)$$

In the non-self-similar problem there are two independent dimensionless variables,

$$\xi = \frac{r}{k(Et^2/\rho_0)^{1/5}} \quad \text{and} \quad \tau = \frac{R_0}{(Et^2/\rho_0)^{1/5}}, \qquad (13.17)$$

on which solutions of gas dynamics equations depend.

To find typical properties of the gas motion in this case we consider propagation of a strong spherical shock wave in the so-called "snow plow" approximation. We assume the shock wave to have the radius $R(t)$, pressure $P(t) \approx \rho_0 T(t)$, and temperature $T(t)$ at the front. For the temperature, we have $T(t) \approx E(t)/M(t)$, where $M(t)$ and $E(t)$ are the mass of gas and energy contained in the region encompassed by he shock wave. Using the relationships for the mass and pressure

$$M = \frac{4\pi}{3}\rho_0 R^3, \qquad P = a\rho_0\frac{E}{M}, \qquad (13.18)$$

we write equations of the shock wave motion as

$$\dot{R} = bP\rho_0^{1/2}, \qquad \dot{E} = -cR^2 T^{3/2}, \qquad (13.19)$$

where a, b, and c are constants. In the case $\gamma_1 = \gamma$ the coefficient c vanishes, $E =$constant, and we have self-similarity of the first type. When $\gamma_1 \neq \gamma$ we obtain

$$\dot{R} = fE^{1/2}R^{-3/2}, \qquad \dot{E} = -gE^{3/2}R^{-5/2}, \qquad (13.20)$$

and

$$\frac{dE}{dR} = -\kappa\frac{E}{R}, \qquad (13.21)$$

where f, g, and κ are constants. The dimensionless exponent κ depends on the dimensionless quantities γ_1 and γ with $\kappa \sim \gamma - \gamma_1$ for $\gamma \ll \gamma_1$. From this we find the power-law relations

$$E \sim R^{-\kappa}, \qquad R \sim t^{2/(\kappa+5)}. \qquad (13.22)$$

The determination of the parameter κ is related to the nonlinear eigenvalue problem for the system of gas-dynamics equations and boundary conditions at the shock wave front.

13.3   Π-theorem

One of the approaches used in current literature to obtain self-similar solutions is to perform a dimension analysis. The dimension analysis, as it is well known, is based on the Π-theorem. This theorem states that the relation among $n + 1$ dimensional quantities $a_1, a_2, \ldots a_n$,

$$a = f(a_1, \ldots, a_k, a_{k+1}, \ldots, a_n) \qquad (13.23)$$

can be expressed in the form

$$\Pi = F(\Pi_1, \ldots, \Pi_{n-k}), \qquad (13.24)$$

where

$$\Pi = \frac{a}{a_1^{\alpha} \ldots a_k^{\kappa}}, \qquad \Pi_1 = \frac{a_{k+1}}{a_1^{\alpha_1} \ldots a_k^{\kappa_1}}, \ldots \ldots, \qquad \Pi_{n-k} = \frac{a_n}{a_1^{\alpha_{n-k}} \ldots a_k^{\kappa_{n-k}}}.$$
$$(13.25)$$

The quantities $a_1, \ldots, a_k$ have independent dimensions. The conclusion that a particular governing parameter, a_i, is unimportant is usually drawn on the basis of the estimate of the value of the corresponding dimensionless parameter Π_i.

If Π_i is either "small" or "large" we replace it in equation (13.23) on zero or infinity respectively. As a result, instead of a function of $n - k$ arguments, F, we obtain a function of $n - k - 1$ variables, Φ. Here Φ is a limit of a function F when Π_i tends to either zero or infinity:

$$\lim_{\Pi_i \to 0, \infty} F = \Phi(\Pi_1, \ldots, \Pi_{i-1}, \Pi_{i+1}, \ldots, \Pi_{n-k}). \qquad (13.26)$$

Example. We analyze the expression for the force, F, acting on the body that moves in a fluid. The sizes of the body are R and L, respectively.

The body moves with the velocity V in the fluid with the density ρ. If one takes into account the viscosity ν, the dimensionless parameters are R/L and the Reynolds number $Re = VL/\nu$. Expression (13.24) in this case can be written as

$$\frac{F}{\rho v^2 S} = \Phi\left(\frac{R}{L}, Re, \ldots\right) \qquad (13.27)$$

with $S = \pi R^2$.

•

However, the conditions under which the limit (13.26) exists in general can not be fulfilled. The function $F(\Pi_i)$, when Π_i tends to zero or infinity, could not have a limit that exists and does not equal zero or infinity. In this case, the preservation of the governing parameter a_{k+i} is essential, no matter how small or large the corresponding dimensionless parameter Π_i is.

Now we consider special cases.

i. We suppose that the limit of a function $F(\Pi_1, \ldots, \Pi_i, \ldots, \Pi_{n-k})$ as $\Pi_i \to 0$ or $\Pi_i \to \infty$ is either zero or infinity, but there is a number α such that there exists the limit

$$\lim_{\Pi_i \to 0, \infty} \Pi_i^{-\alpha} F = \Phi(\Pi_1, \ldots, \Pi_{i-1}, \Pi_{i+1}, \ldots, \Pi_{n-k}). \qquad (13.28)$$

ii. We suppose that the limit of a function $F(\Pi_1, \ldots, \Pi_i, \ldots, \Pi_{n-k})$ as $\Pi_i \to 0$ or $\Pi_i \to \infty$ and $\Pi_j \to 0$ or $\Pi_j \to \infty$ does not exists, but there are numbers α and β such that there exists a limit

$$\lim_{\Pi_i, \Pi_j \to 0, \infty} \Pi_i^{-\alpha} F = \Phi(\Pi_j / \Pi_i^\beta, \Pi_1, \ldots, \Pi_{i-1}, \Pi_{i+1}, \ldots, \Pi_{j-1}, \Pi_{j+1}, \ldots, \Pi_{n-k}).$$

$$(13.29)$$

iii. Three parameters tend to infinity or zero and so on.

The self-similar solution corresponds to the degenerate problem, and is obtained when some parameter, Π_i, tends to zero or infinity. Simultaneously, this must be the exact solution to the degenerate problem with a fewer number of governing parameters and must be an asymptotic representation of non-self-similar solutions.

If the limit of the function F exists and is finite, then as a result of the limiting process the self-similar solutions of the first type are obtained. For these solutions the exponent α in the expression $\xi = x/t^\alpha$ for a self-similar variable can be found from dimensional consideration, using the datum of the degenerate problem as a base.

Self-similar solutions of the second type owe their existence to the special cases, 1 and 2.

A more rigorous approach to find the self-similar solutions is based on the Lie group theory. The self-similar solutions belong to the class of solutions that is invariant under transformations generated by a group.

13.4 Uniformly Propagating Waves in the Framework of the Burgers and Korteweg-de Vries equations as Self-Similar Solutions.

The solution to the Burgers equation given by the expression in the form of a progressive wave,

$$u(x, t) = u(x - t),$$

$$(13.30)$$

represents also the self-similar solution of the first type. Transforming the variables, x and t, to new variables $s = \exp(x)$ and $\tau = \exp(t)$, we obtain

$$u(s, \tau) = u(\ln(s/\tau^V)) = U(\xi),$$

$$(13.31)$$

with $\xi = s/\tau^V$, where $V = (u_1 + u_2)/2$ is determined by the behavior of the solution in large: by the boundary conditions.

A soliton solution to the KdV equation,

$$u(x, t) = \frac{3\alpha^2}{\cosh^2\left(\frac{1}{2}\alpha(x + \alpha^2 t)\right)}$$

$$(13.32)$$

by changing the variables $s = \exp(x)$ and $\tau = \exp(t)$, can be transformed to

$$u(s, \tau) = \frac{6\alpha^2 (s/\tau^{\alpha^2})^{\alpha/2}}{1 + (s/\tau^{\alpha^2})^\alpha},$$

$$(13.33)$$

which depends on the self-similar variable $\zeta = s/\tau^{\alpha^2}$. Here the index α^2 of self-similar variable ζ is proportional to the discrete eigenvalue of the Schrödinger operator: $\alpha^2 = 2\lambda_n/3$.

We see that these self-similar solutions can not be obtained only from the dimensional analysis.

13.5   Self-similar Solutions of the Gas Dynamics Equations

The gas dynamics equations (13.11)-(13.13) admit 5 symmetry transformations. They are invariant with respect to time translation and to the stretching of the density, time and space scale. In the case $n = 2/(\gamma-1)$, there is a symmetry with respect to projective transformation. This includes the case $\gamma = 5/3$ provided $n = 3$. A projective transformation corresponds to mapping

$$x \rightarrow \frac{x'}{1 - \sigma x'}, \qquad y \rightarrow \frac{y'}{1 - \sigma y'}. \tag{13.34}$$

The scale-invariant solutions in general case can be written as

$$v = \frac{r}{t}V(\xi), \qquad \rho = r^\beta \Phi(\xi), \qquad p = \left(\frac{r}{t}\right)^2 P(\xi), \tag{13.35}$$

with the self-similar variable

$$\xi = r/t^\alpha. \tag{13.36}$$

In the limit $\alpha \rightarrow \infty$ the self-similar solution with power law dependence is reduced to solutions, exponential in time. They correspond to invariance with respect to time translation with scaling of space:

$$v = \exp(\pm t/\tau)V(\xi), \qquad \rho = \exp(\pm\beta t/\tau)\Phi(\xi), \qquad p = \exp(\pm 2t/\tau)P(\xi), \tag{13.37}$$

where

$$\xi = r\exp(\pm t/\tau). \tag{13.38}$$

13.6   Self-Similar Solutions with Uniform Deformation

Once more we consider the equations that describe the motion of a gas of non-interacting particles:

$$\partial_t \rho + \partial_x(\rho v) = 0, \tag{13.39}$$

$$\partial_t v + v\partial_x v = 0. \tag{13.40}$$

By direct inspection we see that the equations have exact solutions which we write in the form

$$\rho(x,t) = \rho_0(t) + \rho_2(t)\, x^2, \tag{13.41}$$

$$v(x,t) = v_0(t) + w(t)\, x. \tag{13.42}$$

From equations (13.39), (13.40) we obtain that the functions $v_0(t)$, $w(t)$, $\rho_0(t)$, and $\rho_2(t)$ obey the equations

$$\dot{\rho}_0 + w\rho_0 = 0, \tag{13.43}$$

$$\dot{\rho}_2 + 3w\rho_2 = 0, \tag{13.44}$$

$$\dot{v}_0 + wv_0 = 0, \tag{13.45}$$

$$\dot{w} + w^2 = 0, \tag{13.46}$$

which are ordinary differential equations.

Now we change the variable w, introducing the function $s(t)$ according to the relationship

$$w = \frac{\dot{s}}{s}. \tag{13.47}$$

As a result, instead of the nonlinear equation (13.46) for $w(t)$ we obtain a linear equation for $s(t)$. It is

$$\ddot{s} = 0. \tag{13.48}$$

The solution to this equation is

$$s(t) = 1 + \dot{s}_0 t, \tag{13.49}$$

where $\dot{s}_0 = w(0)$.

Solutions to equations (13.43), (13.44) and (13.45) are expressed in terms of $s(t)$ as follows:

$$\rho_0(t) = \frac{\rho_0(0)}{s(t)}, \qquad \rho_2(t) = \frac{\rho_2(0)}{s^3(t)}, \qquad v_0(t) = \frac{v_0(0)}{s(t)}. \tag{13.50}$$

Finally, we can write for $\rho(x,t)$ and $v(x,t)$

$$\rho(x,t) = \frac{1}{s(t)}\left(\rho_0(0) + \frac{\rho_2(0)}{s^2(t)}x^2\right) = \frac{1}{(1+w(0)t)}\left(\rho_0(0) + \frac{\rho_2(0)}{(1+w(0)t)^2}x^2\right), \tag{13.51}$$

$$v(x,t) = \frac{1}{s(t)}\left(v_0(0) + w(0)\,x\right) = \frac{v_0(0) + w(0)x}{1 + w(0)t}. \tag{13.52}$$

If the gradient of the velocity at $t = 0$ is positive, $w(0) > 0$, then asymptotically for $t \to \infty$ both $\rho(x,t)$ and $v(x,t)$ tend to zero. When $w(0) < 0$ in a finite time, the solutions given by equations (13.51) and (13.52) become singular. The gradients of both the density and the velocity tend to infinity. This time equals

$$t_{\mathrm{br}} = \frac{1}{w(0)}. \tag{13.53}$$

This is a time of wavebreaking, a time of wave collapse, or a time of the gradient catastrophe development.

What is the meaning of the self-similar solutions with uniform deformation that we have discussed? They are the local approximation of a fluid flow in the vicinity of the points where the gradients of the flow velocity are maximal. •

13.6.1 3D Problem

Now we consider the equation

$$\partial_t \mathbf{v} + (\mathbf{v}\nabla)\mathbf{v} = 0. \tag{13.54}$$

In tensor notation we write this equation as

$$\partial_t u_i + u_j \partial_{x_j} u_i = 0. \tag{13.55}$$

This equation has a solution with uniform deformation

$$v_i(\mathbf{x}, t) = v_i(t) + w_{ij}(t)x_j. \tag{13.56}$$

From equations (13.55) and (13.56) we obtain ordinary differential equations for $u_i(t)$ and $w_{ij}(t)$:

$$\dot{v}_i + w_{ij}v_j = 0, \tag{13.57}$$

$$\dot{w}_{ij} + w_{ik}w_{kj} = 0. \tag{13.58}$$

To solve the system (13.57) and (13.58) we perform a transformation of variables from the Euler coordinates x_i to the Lagrange coordinates x_i^0 with the formula

$$x_i = m_i(t) + M_{ij}(t)x_j^0. \tag{13.59}$$

Here $M_{ij}(t)$ is a matrix (3×3); this is a matrix of deformation. As we know, in the Lagrange coordinates

$$v_i(\mathbf{x}, t) = \partial_t x_i. \tag{13.60}$$

Using expressions (13.56), (13.59), and (13.60) we obtain

$$v_i(\mathbf{x}, t) = v_i(t) + w_{ij}(t)x_j =$$

$$= v_i(t) + w_{ij}(t)m_j(t) + w_{ik}(t)M_{kj}(t)x_j^0 = \dot{m}_i(t) + \dot{M}_{ij}(t)x_j^0. \tag{13.61}$$

From this relationship it follows that

$$w_{ij}(t) = \dot{M}_{ik}(t)M_{kj}^{-1}(t), \tag{13.62}$$

and

$$v_i(t) = \dot{m}_i(t) - \dot{M}_{ik}(t)M_{kl}^{-1}(t)m_l(t). \tag{13.63}$$

Here M_{ij}^{-1} is the inverse matrix to the matrix M_{ij}.

Substituting $w_{ij} = \dot{M}_{ik}M_{kj}^{-1}$ into equation (13.58) we find that the matrix of deformation obeys equation

$$\ddot{M}_{kj} = 0 \tag{13.64}$$

with initial conditions:

$$M_{ij}(0) = \delta_{ij}, \qquad \dot{M}_{ij}(0) = w_{ij}(0). \tag{13.65}$$

The solution to equation (13.64) is

$$M_{ij}(t) = \delta_{ij} + w_{ij}(0)t. \tag{13.66}$$

The catastrophe corresponds to the situation when the determinant of the matrix M_{ij},

$$D = \det(M_{ij}), \tag{13.67}$$

vanishes. That follows from both the relationship (13.62) and the dependence of the density on the coordinates and time given by the solution to the continuity equation

$$\partial_t \rho + \operatorname{div}(\rho \mathbf{v}) = 0. \tag{13.68}$$

In the Lagrange coordinates this equation has the solution

$$\rho(\mathbf{x}, t) = \rho(\mathbf{x}, 0) \det\left(\frac{\partial x_i^0}{\partial x_j}\right). \tag{13.69}$$

In our case, when the matrix M_{ij} does not depend on coordinates, the Jacobian, $\det\left(\partial x_i^0/\partial x_j\right)$, is equal to $D^{-1} = 1/\det(M_{ij})$. When D tends to zero, both gradients of the velocity field w_{ij} and the density ρ tend to infinity.

Example. Let us consider the initial matrix of the fluid velocity gradients of the form

$$w_{ij}(0) = \operatorname{diag}(a, b, c). \tag{13.70}$$

That means the matrix $w_{ij}(0)$ is

$$w_{ij}(0) = \begin{pmatrix} a & 0 & 0 \\ 0 & b & 0 \\ 0 & 0 & c \end{pmatrix}. \tag{13.71}$$

In this case, the matrix $M_{ij}(t)$, according to relationship (13.66), is equal to

$$M_{ij}(t) = \begin{pmatrix} 1+at & 0 & 0 \\ 0 & 1+bt & 0 \\ 0 & 0 & 1+ct \end{pmatrix}. \tag{13.72}$$

Its determinant is

$$D(t) = (1+at)(1+bt)(1+ct). \tag{13.73}$$

Singularity occurs when either $t = -1/a$, $t = -1/b$, or $t = -1/c$. Singularity appears in a point when all the values of a, b and c are equal to each other and are negative. It appears on a line for two equal and negative values among a, b and c, and singularity appears on a surface when just one value among a, b and c is negative. We see that the generic case corresponds to the situation when just one value among a, b and c is negative. That means that in the generic case the singularity develops on the surface.

•

References

1. V. I. Arnol'd, Mathematical Methods of Classical Mechanics (Nauka, Moscow,1971).
2. G.I. Barenblatt, Ya. B. Zel'dovich, *Ann. Rev. of Fluid Mech.* **4**, 285 (1972).
3. S.V.Bulanov, P.V.Sasorov, *Sov. J. Plasma Phys.* **4**, 418 (1978).
4. S.V.Bulanov, P.V.Sasorov, *Sov. Phys. JETP* **59**, 279 (1984).
5. S.V.Bulanov, M.A.Ol'shanetskij, *Physics Letters A* **100**, 35 (1984).
6. S.V.Bulanov, A.S.Sakharov, *JETP Lett.* **44**, 543 (1986).
7. S.V.Bulanov, F. Cap, *Soviet Astronomy J.* **32**, 436 (1988).
8. S.V.Bulanov, F.Pegoraro, A.S.Sakharov, *Physics of Fluids B*, **4**, 1992 (1992).
9. S.V.Bulanov, N.M.Naumova, *Physica Scripta*, **T63** , 190 (1996).
10. S.V.Bulanov, M.Lontano, T.Zh.Esirkepov, F.Pegoraro, A.M.Pukhov, *Phys. Rev. Lett.* **76**, 3562 (1996).
11. S.V.Bulanov, T.Zh.Esirkepov, M.Lontano, F.Pegoraro, *Plasma Physics Reports* **23,** 660 (1997).
12. S.V. Bulanov, F.Pegoraro, A.M.Pukhov, A.S.Sakharov, *Phys. Rev. Lett.*, **78**, 4205 (1997).
13. J. R. Cary, D.F.Escande D. F., J.L. Tennyson, *Phys. Rev. A.* **34**, 4256 (1986).
14. C.S. Gardner, J.M. Green, M.D. Kruskal, and R.M. Miura, *Phys. Rev. Lett.* **19**, 1095 (1967).
15. J.H.Hannay, *J. Phys. A* **18**, 221 (1985).
16. H. Lamb, Hydrodynamics (Cambridge University Press, Cambridge, 1932).
17. L.D.Landau, E.M.Lifshits, Mechanics (Nauka, Moscow, 1973).
18. L.D.Landau, E.M.Lifshits, Hydrodynamics (Nauka, Moscow, 1973).
19. A. Lichtenberg, M. Liberman, Regular and stochastic dynamics. 1984.
20. P. D. Lax, *Com. Pure Appl. Math.* **21**, 467 (1968).
21. V.K.Melnikov, *Soviet Physics - Doklady* **144**, 747 (1962); ibid., **148**, 1257, (1963).
22. V. A. Marchenko, The Sturm-Liouville Operators and their Applications (Naukova Dumka, Kiev, 1977).
23. A.I. Neishtadt, *Sov. J. Phys. Plas.* **12**, 992 (1986).
24. E. Ott, *Phys. Rev. Lett.* **29**, 1429 (1972).
25. V. I. Petviashvili, and O. M. Pokhotelov, Solitary Waves in Plasmas and in the Atmosphere (Gordon and Breach Science Publishers, New York, 1992).

26. T. Poston, Y. Steward, Catastrophe Theory and its Applications, (Pitman, London, 1978).

27. A. V. Timofeev, *Sov. J. JETP* **75**, 1303 (1978).

28. G.B. Whitham, Linear and Nonlinear Waves (Wiley. N.Y., 1974).

29. V. E. Zakharov, A. B. Shabat, *JETP* **71**, 203 (1976).

30. G. M. Zaslavskij, R. Z. Sagdeev, Introduction to nonlinear physics (Nauka, Moscow,1988).

Elenco dei volumi della collana
"Appunti"
pubblicati dall'Anno Accademico 1994/95

GIUSEPPE BERTIN (a cura di), *Seminario di Astrofisica,* 1995.

EDOARDO VESENTINI, *Introduction to continuous semigroups,* 1996.

LUIGI AMBROSIO, *Corso introduttivo alla Teoria Geometrica della Misura ed alle Superfici Minime,* 1997.

CARLO PETRONIO, *A Theorem of Eliashberg and Thurston on Foliations and Contact Structures,* 1997.

MARIO TOSI, *Introduction to Statistical Mechanics and Thermodynamics,* 1997.

MARIO TOSI, *Introduction to the Theory of Many-Body Systems,* 1997.

PAOLO ALUFFI (a cura di), *Quantum cohomology at the Mittag-Leffler Institute,* 1997.

GILBERTO BINI, CORRADO DE CONCINI, MARZIA POLITO, CLAUDIO PROCESI, *On the Work of Givental Relative to Mirror Symmetry,* 1998

GIUSEPPE DA PRATO, *Introduction to differential stochastic equations,* 1998

HERBERT CLEMENS, *Introduction to Hodge Theory,* 1998

HUYÊN PHAM, *Imperfections de Marchés et Méthodes d'Evaluation et Couverture d'Options,* 1998

MARCO MANETTI, *Corso introduttivo alla Geometria Algebrica,* 1998

AA.VV., *Seminari di Geometria Algebrica 1998-1999,* 1999

ALESSANDRA LUNARDI, *Interpolation Theory,* 1999

RENATA SCOGNAMILLO, *Rappresentazioni dei gruppi finiti e loro caratteri,* 1999

SERGIO RODRIGUEZ, *Symmetry in Physics,* 1999

F. STROCCHI, *Symmetry Breaking in Classical Systems and Nonlinear Functional Analysis,* 1999

LUIGI AMBROSIO e PAOLO TILLI, *Selected Topics on "Analysis in Metric Spaces",* 2000

ANDREA C.G. MENNUCCI, SANJOY K. MITTER, *Probabilità ed informazione,* 2000

SERGEI VLADIMIROVICH BULANOV, *Lectures on Nonlinear Physics,* 2000

"CompoMat" Loc. Braccone, 02040 Configni (RI), Italy
Finito di stampare dalla Nuova Grafica 86 nel dicembre 2000